Mathematics for Chemists

Mathematics for Chemists

P. G. Francis

Department of Chemistry,
University of Hull

LONDON   NEW YORK
Chapman and Hall

*First published 1984
by Chapman and Hall Ltd
11 New Fetter Lane, London EC4P 4EE*

*Published in the USA by
Chapman and Hall
733 Third Avenue, New York NY 10017*

© 1984 P. G. Francis

Printed in Great Britain by J. W. Arrowsmith Ltd., Bristol

ISBN 0 412 24980 4 (cased)
ISBN 0 412 24990 1 (Science Paperback)

British Library Cataloguing in Publication Data

Francis, P.G.
  Mathematics for chemists.
  1. Chemistry—Mathematics
  I. Title
  510'.2454   QD42

  ISBN 0–412–24980–4
  ISBN 0–412–24990–1 Pbk

Library of Congress Cataloging in Publication Data

Francis, P. G. (Patrick Geoffrey), 1929–
  Mathematics for chemists.

  Bibliography: p.
  Includes index
  1. Mathematics—1961   2. Chemistry—
Mathematics. I. Title.
  QA37.2.F73 1984 510'.2454 84–1802
  ISBN 0–412–24980–4
  ISBN 0–412–24990–1

Contents

Preface

This text is concerned with those aspects of mathematics that are necessary for first-degree students of chemistry. It is written from the point of view that an element of mathematical rigour is essential for a proper appreciation of the scope and limitations of mathematical methods, and that the connection between physical principles and their mathematical formulation requires at least as much study as the mathematical principles themselves. It is written with chemistry students particularly in mind because that subject provides a point of view that differs in some respects from that of students of other scientific disciplines. Chemists in particular need insight into three-dimensional geometry and an appreciation of problems involving many variables. It is also a subject that draws particular benefit from having available two rigorous disciplines, those of mathematics and of thermodynamics. The benefit of rigour is that it provides a degree of certainty which is valuable in a subject of such complexity as is provided by the behaviour of real chemical systems. As an experimental science, we attempt in chemistry to understand and to predict behaviour by combining precise experimental measurement with such rigorous theory as may be at the time available; these seldom provide a complete picture but do enable areas of uncertainty to be identified. Mathematical rigour has been provided for us by generations of professional mathematicians, who continue to give support and advice both in the application of established techniques and in the development of new approaches. Experimental scientists have added to this a rigour of a different kind, which is based upon a small number of premises, or axioms, deduced from seemingly disconnected experimental observations. These provide the experimental laws of thermodynamics and quantum theory whose justification lies in the absence of disagreement with experiment. When these are expressed in mathematical terms we can add mathematical rigour to produce theoretical structures in which only the basic premises, often corresponding to a simplified physical model, leave room for uncertainty. These considerations are reflected in this text by frequent reference to physical principles, by the inclusion of a number of mathematical proofs and by the inclusion of a discussion of the treatment of experimental data.

It is assumed that the reader has previous mathematical knowledge extending to the elementary applications of calculus. The subject matter is developed during each chapter, so that an initial reading of only the earlier sections in each chapter is possible. The later sections contain brief outlines of the principles and scope of more advanced techniques and are intended to point the way to more specialized texts.

The examples are, in the main, given with their solutions; this is in order that the text may be used as a reference source, for which purpose worked examples are more valuable than exercises. Readers are strongly recommended to attempt their own solutions to these examples before consulting the one that is given.

A proper acknowledgement of the many sources from which the author has drawn cannot be justly be made for a text in which the material was largely developed during the last two- or three-hundred years. It is a subject on which many admirable texts have been written, the only justification for adding to their number being changes in emphasis.

CHAPTER 1

Algebraic and geometrical methods

1.1 Natural numbers

Quantitative measurements in science are usually made in terms of appropriate, but arbitrary, units (Section 1.2), so that the numerical values obtained depend on the size of the units and are not significant in themselves. Only when there is a natural unit, such as the gas constant R, do numerical values assume absolute significance; for example, the heat capacity at constant volume of an ideal monatomic gas at a high enough temperature is $3R/2$, where the number 3 corresponds to the three degrees of freedom of the particles.

There are some pure numbers, however, that arise naturally, the ones most commonly met in science being π and e. These can be regarded as the mathematical equivalent of natural units and occur so often in scientific formulae and equations that we consider how they originate.

Our first natural number, π, could be defined as the ratio of circumference to diameter for a circle. This ratio is a pure number that is independent of the size of the circle, and it is also used as the natural measure of angle. The definition of angle is

$$\text{angle} = (\text{length of arc})/\text{radius}, \tag{1.1a}$$

and for a given angle, this has a numerical value that is independent of any conventional scale of measurement. The circumference of a complete circle of radius r being $2\pi r$, the angle corresponding to a full rotation is $2\pi r/r = 2\pi$ natural units of angle, this unit being called the radian. We also have the conventional definition of such a full rotation as 360 degrees (360°), so that

$$\pi \text{ radian} = 180°.$$

The definition of angle in two dimensions as (arc length)/radius can be extended into three dimensions; if we take a conical segment from a sphere we can define a three-dimensional solid angle as

solid angle = (area of segment of spherical surface)/(radius)^2.

(1.1b)

For a complete sphere of surface area $4\pi r^2$,

$$\text{complete solid angle} = 4\pi r^2/r^2 = 4\pi,$$

and since this expression does not contain r it is also a pure number that is independent of the size of the sphere. The unit of solid angle is called a steradian. It is a useful geometrical concept that (area)/(radius)^2 is constant for a given solid angle, and this is used in Section 5.7.

Our second natural number, e, arises in a different way. It is useful to divide the mathematical methods used in science into two categories, geometrical methods and analytical methods. The above definition of π is geometrical and belongs in the field of diagrams and models. The second category is algebraic and numerical, and does not depend on our ability to construct models and diagrams. Analytical methods are developed by strictly logical deductions from first principles, called axioms. This is generally the more powerful of the two mathematical methods and, as such, is the one preferred by mathematicians and is often the one needed to solve the more difficult problems. Geometrical methods, on the other hand, often seem simpler but are sometimes deceptively so. The wise rule is to use geometrical methods with care and to defer to analytical methods when in doubt.

This division into geometrical methods and analytical methods is arbitrary since they are parts of the same whole and a combination of the two is frequently used in practice. Sometimes the connection between the two methods is not obvious, an example being provided by the trigonometric functions $\sin\theta$, $\cos\theta$, $\tan\theta$, ... The geometrical definition of $\sin\theta$ is

$$\sin\theta = \frac{\text{opposite side}}{\text{hypotenuse}} \quad \text{in a right-angled triangle,} \quad (1.2)$$

whereas the analytical definition is

$$\sin\theta = \theta - \frac{\theta^3}{3!} + \frac{\theta^5}{5!} - \frac{\theta^7}{7!} + \dots . \quad (1.3)$$

We can see that these two definitions are the same in a particular case by taking the angle θ to be, say, $90° = \pi/2$ radian. Then $\sin \theta = 1$ by the geometrical definition, and if we put $\theta = \pi/2 = 1.5708$ into the series (1.3) it is easily shown that successive terms become rapidly smaller and smaller, and that the sum tends towards a limit of 1.000.

The fact that the two definitions of $\sin \theta$ are always the same can be proved only by means of extensive analytical argument. One way is to use the definition of $\cos \theta$ as the series

$$\cos \theta = 1 - \frac{\theta^2}{2!} + \frac{\theta^4}{4!} - \frac{\theta^6}{6!} + \dots \qquad (1.4)$$

This is then the derivative (Chapter 2) of $\sin \theta$. The derivative is used to define the slope of the tangent to a curve, and the angle between two straight lines can be obtained as the inverse of the cosine series. In this way, geometry can be developed by the analytical approach leading eventually to showing that (1.2) and (1.3) are equivalent to each other.

As mentioned above, our second natural number, e, belongs in analytical methods. It is defined by the series

$$e^x = 1 + x + \frac{x^2}{2!} + \frac{x^3}{3!} + \dots, \qquad (1.5)$$

so that when $x = 1$,

$$e = 1 + 1 + \frac{1}{2!} + \frac{1}{3!} + \dots = 2.71828\dots\dots \qquad (1.6)$$

This arises naturally in science because e^x, and only e^x, gives exactly the same quantity when it is differentiated (Chapter 2). This corresponds to the quite common situation where the rate of change of a quantity is proportional to the quantity itself, as in first-order reaction kinetics. More fundamentally, exponential relations arise as a result of the simple probability that determines the Boltzmann distribution which underlies many physical phenomena (Section 2.8).

Putting $x = 0$ in (1.5) gives $e^0 = 1$. This is a particular case of the general rule that any quantity raised to the power of zero is unity, which follows from the laws of indices:

$$a^x a^y = a^{(x+y)} \qquad \text{and} \qquad a^{-x} = 1/a^x,$$

so that

$$a^{(x-y)} = a^x/a^y,$$

and when $x = y$,

$$a^0 = 1.$$

1.2   Units and dimensional analysis

The units used in the SI system are given in the appendix together with numerical values of the fundamental physical constants and conversion factors. Notice that the symbols for units are written only in the singular, or they cannot be cleanly cancelled and we would also have absurdities such as 0.99 g but 1.01 gms. This need not conflict with colloquial usage; we may speak of a temperature difference of 10 kelvins but this is written as 10 K.

Algebraic symbols are used to denote physical quantities, such as p for pressure; the symbol denotes the quantity, not the units. A particular value of the quantity is then the product of a pure number and its units, and this product follows the normal rules of algebra. Thus 2 atm is the product of the pure number 2 and the unit atm, so that if $p = 2$ atm we have p/atm = 2, or a quantity divided by its units is a pure number. An important particular case is when taking logarithms; since the definition of a logarithm only applies to pure numbers, the value of any quantity must first be divided by its units, such as $\ln(p/\text{atm})$. The value obtained will then depend on the units being used since, for example,

$$\ln(p/\text{atm}) = \ln(p/101.325\,\text{kPa})$$
$$= \ln(p/\text{kPa}) - \ln(101.325)$$
$$= \ln(p/\text{kPa}) - 4.618.$$

The dimensions of a physical quantity are defined as the appropriate combination of powers of the fundamentals mass (M), length (L) and time (T); thus velocity has dimensions LT^{-1} and force (mass × acceleration) has dimensions MLT^{-2}. The dimensions are precisely analogous to the SI units of kilogram, metre and second, so that force has units of kg m s^{-2}.

Dimensional analysis is a method of checking and predicting relations between physical quantities based simply on the principle that the dimensions must balance in any equation. A standard example of the technique is to predict the form for Stokes's law for the drag on a sphere moving in a viscous fluid. Since

$$\text{force} = \text{mass} \times \text{acceleration}$$

it has dimensions MLT^{-2}. The viscosity η of a fluid is defined by

$$\eta = \frac{\text{force per unit area}}{\text{velocity gradient}}$$
$$= \frac{\text{force}}{\text{area}} \times \frac{\text{distance}}{\text{velocity}}$$

with dimensions

$$\frac{M\!\!\!/L}{T^{\not{2}}} \times \frac{1}{L\!\!\!/^2} \times L\!\!\!/ \times \frac{T\!\!\!/}{L} = ML^{-1}T^{-1}.$$

We then assume that the drag (resisting force) on the sphere will depend on the size of the sphere (radius a), its velocity (v) and on the viscosity of the fluid. Dependence on temperature will be taken care of through change in the viscosity of the fluid. The relation that is assumed is written as

$$\text{force} = ka^{\alpha}v^{\beta}\eta^{\gamma}$$

and dimensional analysis will enable us to find the numbers α, β and γ. The assumed relation is rewritten in terms of the dimensions (or units) of the quantities, thus

$$MLT^{-2} = L^{\alpha}(LT^{-1})^{\beta}(ML^{-1}T^{-1})^{\gamma}.$$

Finally, we equate powers of each of the primary dimensions M, L and T to obtain

$$\text{for } M, \qquad \gamma = 1,$$
$$\text{for } L, \qquad \alpha + \beta - \gamma = 1,$$
$$\text{for } T, \qquad -\beta - \gamma = -2,$$

so that $\alpha = \beta = \gamma = 1$ and the required equation is

$$\text{force} = ka\eta v.$$

Physical, rather than dimensional, arguments are needed to show that the proportionality constant k is 6π.

The same principle can be applied when using the practical SI units of kilogram (kg), metre (m) and second (s), so that in any equation connecting physical variables the units must cancel on the two sides of the equation. This has two valuable and practical uses: unless the units cancel, the equation being used is wrong; and when we wish to change to other units, we can do so by a simple algebraic method called quantity calculus. The latter is based on the principle that multiplication by unity leaves any quantity unchanged, and we can construct 'unity brackets' by writing, as a fraction, the new units over the old using appropriate conversion factors. As a simple example, if we write (1 week/7 days) this has the value unity. To convert, say, 42 days into weeks we multiply by the appropriate unity bracket and cancel the units:

$$42 \text{ days} = 42 \text{ days} \times \frac{1 \text{ week}}{7 \text{ days}} = 6 \text{ weeks}.$$

Example 1.1

The gas constant R in the equation $pV = nRT$ has the units needed for the equation

$$R = \frac{pV}{nT}$$

to have the same units on both sides. Thus if the pressure p is in atmospheres (atm), volume V in dm^3 (also called litres), and temperature T in kelvins (K), and amount of substance, n in moles (mol)

$$R = 0.08205 \text{ litre atm mol}^{-1} \text{ K}^{-1}.$$

If we wish to use SI units instead, we make use of unity brackets based on the conversion factors

$$1 \text{ atm} = 101.325 \text{ kPa},$$
$$1 \text{ J} \quad = 1 \text{ N m},$$
$$1 \text{ dm}^3 = 10^{-3} \text{ m}^3,$$
$$1 \text{ Pa} \quad = 1 \text{ N m}^{-2},$$

so that, using four significant figures,

$$R =$$

$$\frac{0.082057 \, dm^3 \, atm}{\text{mol K}} \times \frac{10^{-3} \, m^3}{dm^3} \times \frac{101.325 \times 10^3 \, Pa}{atm} \times \frac{N}{Pa \, m^2} \times \frac{J}{N \, m}$$

$$= 8.3144 \text{ J mol}^{-1} \text{ K}^{-1}.$$

Example 1.2

Show that Planck's constant has the dimensions of momentum × length.

From the appendix,

$$h = 6.624 \times 10^{-34} \text{ J s}.$$

The units of momentum × length are those of mass × velocity × length. The unity brackets are based on the conversion factors

$$1 \text{ J} = 1 \text{ N m},$$
$$1 \text{ N} = 1 \text{ kg m s}^{-2},$$

so that the units of momentum × length are

$$\text{kg} \times \frac{m}{s} \times m = \frac{kg \, m^2}{s} \times \frac{N \, s^2}{kg \, m} \times \frac{J}{N \, m} = J \, s$$

Example 1.3

Suppose that we are given that the spacing between the lines in the microwave spectrum of HCl is 20.7 cm^{-1} and that we need to calculate the moment of inertia I, but can remember only that the spacing is $2B$ where B is either $h^2/8\pi^2 I$ or $h/8\pi^2 Ic$. This difficulty can be resolved by adopting the strongly recommended rule that we substitute into formulae not just numerical values but the units as well.

We have that either

$$I = \frac{h^2}{8\pi^2} \times \frac{\text{cm}}{10.35} \quad \text{or} \quad I = \frac{h}{8\pi^2 c} \times \frac{\text{cm}}{10.35},$$

where $h = 6.626 \times 10^{-34}$ J s and $c = 2.998 \times 10^8$ m s^{-1}, so that either

$$I = \frac{(6.626 \times 10^{-34})^2 \, \cancel{J^2} \cancel{s^2} \, \text{cm}}{8\pi^2 \times 10.35} \times \frac{\cancel{N^2} \, \text{m}^2}{\cancel{J^2}} \times \frac{\text{kg}^2 \, \text{m}^2}{\cancel{N^2} \cancel{s^{4}} \cancel{2}} \times \frac{\text{m}}{10^2 \, \text{cm}}$$

giving units of $\text{kg}^2 \, \text{m}^3 \, \text{s}^{-2}$, or

$$I = \frac{6.626 \times 10^{-34} \, \cancel{J} \cancel{s} \cancel{cm} \cancel{s}}{8\pi^2 \times 2.998 \times 10^8 \, \cancel{m} \times 10.35} \times \frac{\cancel{N} \, \text{m}}{\cancel{J}} \times \frac{\text{kg} \, \text{m}}{\cancel{N} \cancel{s^2}} \times \frac{\cancel{m}}{10^2 \, \cancel{cm}}$$

$$= 2.70 \times 10^{-47} \, \text{kg} \, \text{m}^2.$$

The second expression gives the correct units for the moment of inertia.

Example 1.4

Show that a diver at a depth of 66 ft of sea water (density 1.03 g cm^{-3}) is under a pressure of about 3 atm.

The pressure exerted by a column of liquid of density ρ and height h is given by

$$p = \rho g h = \frac{1.03 \, \text{g}}{\cancel{cm^3}} \times \frac{9.81 \, \cancel{m}}{\cancel{s^2}} \times 66 \, \cancel{ft} \times \frac{\text{kg}}{10^3 \, \cancel{g}} \times \frac{\cancel{N} \cancel{s^2}}{\text{kg} \, \cancel{m}} \times \frac{12 \, \cancel{in}}{\cancel{ft}}$$

$$\times \frac{2.54 \, \cancel{cm}}{\cancel{in}} \times \frac{10^4 \, \cancel{cm^2}}{\cancel{m^2}} \times \frac{\text{Pa} \, \cancel{m^2}}{\cancel{N}}$$

$$= 2.033 \times 10^5 \, \text{Pa} = 203.3 \, \text{kPa}.$$

Thus the water exerts a pressure of about 2×101 kPa (2 atm), and so the total pressure on the diver is about 3 atm.

1.3   Functional notation

When the value of a quantity z depends on the values of one or more quantities $x, y, \ldots$, we call $z, x, y, \ldots$ variables and show z to be a function of $x, y, \ldots$ by writing either

$$z = z(x, y, \ldots),\tag{1.7}$$

or

$$z = f(x, y, \ldots).\tag{1.8}$$

The first form, equation (1.7), is used to show only that a dependence exists, whereas the second labels a particular function as $f(x, y, \ldots)$, and this would not be the same function as, say, $g(x, y, \ldots)$.

Variables are usually denoted by letters from the end of the Greek or Roman alphabets, and constants are usually denoted by letters from the beginning of the alphabets, but this is a useful guideline rather than a universal rule.

We denote the value of a function at particular values of the variables by $f(a, b)$, which is the value of $f(x, y)$ when $x = a$ and $y = b$. Thus the enthalpy H of a substance, which depends on the temperature T and the pressure p, is shown as $H(T, p)$, and at the particular values of the variables $T = 298$ K and $p = 1$ atm we write $H(298$ K, 1 atm).

1.4   Quadratic and higher-order equations

An algebraic equation in a single variable, say x, is said to be of degree n when the highest power of x in the equation is x^n after negative or fractional powers have been removed. Thus the equation

$$x - \frac{2}{x} = 1$$

must first be multiplied through by x to give

$$x^2 - x - 2 = 0\tag{1.9}$$

and so it is a second-degree (or quadratic) equation. An equation that contains x^3 is third degree (or cubic), and so on.

Such equations will be satisfied by particular values of x called the roots of the equation, and an equation of degree n has n roots. An equation is solved by finding the roots by any convenient method. In general, if $x = a$ is a root of the equation $f(x) = 0$, then $(x - a)$ is a factor, which means that the equation can be written as $(x - a)g(x) = 0$. The simplest solution will be when the factors can be seen by inspection. Thus equation (1.9) gives

$$(x-2)(x+1) = 0.$$

This will be satisfied if either $(x-2) = 0$ or $(x+1) = 0$, giving $x = 2$ and $x = -1$ as the roots of the equation.

The factors of a difference between two squares are particularly useful, as

$$x^2 - 4 = (x+2)(x-2).$$

The roots of a cubic equation such as

$$x^3 - 2x^2 - x + 2 = 0 \qquad (1.10)$$

may be found by trial and error. By assuming small whole-number values of x we find that $x = 1$ satisfies the equation; this means that $(x-1)$ is a factor. Long division of the expression by $(x-1)$ gives the other, quadratic, factor, which in this case is the left-hand side of (1.9), so that the roots of (1.10) are $x = 1$, $x = 2$ and $x = -1$.

A quadratic equation that cannot easily be factorized is solved by using the formula for solution of a quadratic, which is obtained by 'completing the square'. Given the equation

$$ax^2 + bx + c = 0$$

we first divide through by a to obtain

$$x^2 + \frac{b}{a}x + \frac{c}{a} = 0.$$

The terms containing the variable x can be obtained by writing

$$\left(x + \frac{b}{2a}\right)^2 = x^2 + \frac{b}{a}x + \frac{b^2}{4a^2}$$

so that the original equation can be written as

$$\left(x + \frac{b}{2a}\right)^2 = \frac{b^2}{4a^2} - \frac{c}{a} = \frac{b^2 - 4ac}{4a^2}.$$

We now take square roots of both sides, recognizing that this introduces ambiguity of sign, so that

$$x = -\frac{b}{2a} \pm \frac{1}{2a}\sqrt{(b^2 - 4ac)}. \qquad (1.11)$$

Corresponding formulae can be written for the solution to a cubic equation, but these are too cumbersome to be useful; it is better to find a numerical factor by trial and error and to use the formula (1.11) for the remaining quadratic factor. Newton's method (Section 2.9.2) can be used in the initial numerical approximation.

Real solutions of a quadratic equation will not exist when $b^2 < 4ac$. We can then use complex numbers to write solutions containing real and imaginary parts as in Section 1.10.

Example 1.5

In physical applications we may expect physical conditions to restrict the choice of roots. Thus the dissociation constant K of a solution of a monobasic weak acid is related to its degree of dissociation α at concentration c by

$$K = \frac{\alpha^2 c}{1 - \alpha}$$

so that

$$c\alpha^2 + K\alpha - K = 0,$$

$$\alpha = -\frac{K}{2c} \pm \frac{1}{2c} \sqrt{(K^2 + 4Kc)}.$$

But K, c and α are all positive, so that only the positive sign before the root has physical meaning (see also Example 2.25).

1.5   Dependent and independent variables

When we say that a variable is independent we mean that it can assume any value. A simple case is if we were concerned with, say, the length of a given bar of metal as a function of temperature. We can fix the temperature at any value we please, and this is an independent variable. Once having fixed the temperature, the length is fixed by the physical properties of the metal, so that length is not independent once the temperature is fixed. This is an example of one dependent and one independent variable.

If we were considering the volume of a sample of gas, this depends not only on temperature but also on pressure. We then have three variables altogether (volume, temperature and pressure) of which two are independent (e.g. temperature and pressure) and one dependent (volume). Notice that we could equally well say that the pressure depends on the volume and the temperature, which would imply that volume and temperature are independent variables and pressure the dependent one. Thus we cannot specify which particular variables are necessarily dependent or independent; what we can specify clearly is that two are independent and the third (whichever one of the three we choose) is dependent.

To determine how many variables in a given problem are independent we need to know how many relevant variables there are (three for our sample of gas) and how many equations exist connecting these variables (one, $pV_m = RT$ for a perfect gas). The number of independent variables is then the total number minus the number of equations ($3 - 1 = 2$ for our sample of gas). The mathematical significance of this is that each equation (simultaneous equation if more than one) can be used to eliminate one variable from the problem; if we have as many simultaneous equations as variables we can solve the equations to obtain numerical values of the variables, meaning that the number of independent variables is then zero. This need to have as many simultaneous equations as variables in order to obtain a unique solution is usefully borne in mind, particularly with problems involving chemical equilibria.

A chemical example of the limitations on the number of independent variables set by the existence of relating equations is the phase rule. Each phase of C components requires $C - 1$ composition variables (the 1 is a relating equation that the sum of the mole fractions is unity), so that P phases require $P(C - 1)$ composition variables. The temperature and pressure are two more variables, giving $P(C - 1) + 2$. The equations relating the variables are equality of chemical potential for each component in each pair of phases, giving $C(P - 1)$ equations. The number of independent variables is then

$$P(C - 1) + 2 - C(P - 1) = C + 2 - P.$$

Alternatively, instead of composition as a variable, we may choose to use the C values of the chemical potential of each component, giving, with temperature and pressure, $C + 2$ variables. In that case we have a Gibbs–Duhem equation relating the chemical potentials in each phase, giving P equations and so, again, $C + 2 - P$ independent variables.

The number of independent variables is also called the number of degrees of freedom of the system, and is the number of variables that must be specified to define the state of a system completely. It is a characteristic of chemical problems that we have more than this minimum number of possible variables, resulting in a freedom of choice which can cause confusion. Thus in the simplest case of a single phase of a pure substance, the two independent variables may be chosen from p, V_m and T; we might choose to use either (p,T) or (V_m,T) as the independent variables and obtain correct results in either case. It often happens that the relevant formulae are simpler in terms of one choice of independent variables than another; for example if the volume is subsequently to be held constant this should be chosen as one of the variables in setting up the necessary equations because the simplest form of the result will then be obtained. Experimental measurements on chemical systems are often conducted at a constant pressure of one atmosphere, so the best choice of independent variables is then (p,T). Theoretical treatments, on the other hand, are often simplest in terms of a fixed volume for the system and the independent variables used are (V_m, T). Thermodynamic properties are expressed in terms of the enthalpy H and Gibbs function G when (p,T) are chosen as the variables; the total energy U and Helmholtz free energy A give the simplest formulae in terms of (V_m, T).

1.6 Graphical methods

Graphical methods of presenting and analysing data are used because they reveal experimental scatter of points, because curvature can be detected or the choice of a best straight line can be made, and because they facilitate interpolation (reading intermediate values between the known points) and extrapolation (the projection of curves or lines beyond the range of the points). In precise work, graphical methods should be regarded only as a prelude to the analytical treatment described in Chapter 7.

The following terms and conventions are used in graph plotting. The vertical or y-axis is the ordinate and the horizontal or x-axis is the abscissa. The scales should be chosen so that the given points fill the graph, apart from short extrapolations, which implies that any straight line will have an angle of slope near to 45°. A long extrapolation is sometimes required, such as to absolute zero of temperature, but this should be calculated, not drawn; similar triangles (Section 1.7.1) are then particularly useful.

The scales on a graph are pure numbers so that only pure number quantities should be plotted. The significance of this is that if the graph were to show, say, length in metres as a function of absolute temperature, the labels on the axes of the graph are length/m and T/K because length divided by metre and temperature divided by kelvin are both pure numbers.

Graphs are always plotted with the dependent variable as ordinate and the independent variable as abscissa. As pointed out in the previous section, there is no mathematical distinction between two variables as to which is the dependent and which the independent variable, so that physical criteria are required. The proper choice is often obvious in that the independent variable is deliberately fixed and the dependent variable is then the result of an experiment. Time, temperature and composition are common independent variables and so should be on the x-axis. When the proper choice is not obvious, precedent should be followed; all graphs of the same quantities will then have the same appearance.

The best graph to plot is a straight line because visual estimation is then most precise and because simple analytical treatment is then available. The equation of a straight line is conventionally written as

$$y = mx + c. \tag{1.12}$$

When $x = 0$, $y = c$, which is the intercept on the y-axis, and when $y = 0$, $x = -c/m$, the intercept on the x-axis. The slope, or gradient, of the graph is m, this being the scaled tangent of the angle between the line and the x-axis and the value of the derivative $\mathrm{d}y/\mathrm{d}x$ (Chapter 2).

The best straight line through a set of points can be fairly precisely estimated by eye using a tight cotton thread so as not to obscure any points. The best line passes through the 'centre of gravity' of the points and the sum of the squares of the deviations of the points from the line should be a minimum, so that large deviations are much more important than small ones. Such a visual choice of line will be subjective, so that when the quality of the data merits it the analytical method of least squares should be used as described in Section 7.8.

When a graph is used to test the agreement between experimental data and a theoretical form of equation, or when numerical values of constant parameters are required, a straight-line graph is much better than a curve. Many equations that are not originally in the linear form of equation (1.12) can be manipulated into that form before the scales of the graph are chosen. Thus the first-order decay of quantity x with time t is given by

$$x = x_0 e^{-kt}.$$

This is converted into linear form by taking natural logarithms of both sides

$$\ln x = -kt + \ln x_0$$

and a straight line should be obtained by plotting $\ln x$ against t (the independent variable, so abscissa or x-axis), with slope $-k$ and intercept $\ln x_0$ on the ordinate (y-axis). When a logarithmic scale is used the numerical values of any physical quantity must be divided by their units as described in Section 1.2; changing these units will change the intercept of a graph but will not affect the slope.

1.7   Some geometrical methods

1.7.1   Similar triangles

If two triangles have corresponding angles equal, then they are said to be similar and the lengths of corresponding sides are all in the same ratio to each other. This is a particularly useful geometrical theorem.

In Fig. 1.1 a straight line passing through points A and B fits experimental data plotted on paper shown in broken outline. The intercepts c and b would be most easily found by fitting the two points to $y = mx + c$ by $m = BN/AN$ and then $c = y_A - mx_A$. Then by similar triangles $c/(-b) = BN/AN = m$, hence $b = -c/m$ as before.

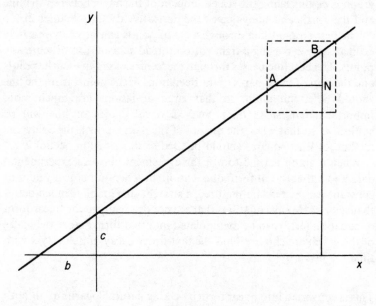

Fig. 1.1

1.7.2   Triangular graph paper

In an equilateral triangle, the sum of the perpendicular distances from any point onto the sides is equal to the height of the triangle. This is used to represent the composition of three component mixtures on two-dimensional paper.

In Fig. 1.2 point O is anywhere within the equilateral triangle ABC of side a and height h. The sum of the perpendiculars $OP + OQ + OR$ can be seen to be equal to h by noting that the total area of the triangle is the sum of the areas of AOB, BOC and COA, each area being $(a/2)$ times the corresponding height.

If a scale is chosen so that the height of the triangle is either unity (in mole fraction) or 100 (in percentage composition), the amount of each of three components can be plotted from zero on a side to 1 (or 100) at the opposite apex. Triangular graph paper is available on which lines parallel to the sides are drawn so that such scales can be used. In Fig. 1.2, the point O is where the percentage of A is QO, that of B is RO and of C is PO; the scale is such that h is 100. The line BX in the figure is the locus of points that would be obtained by adding pure B to a mixture of composition X of A and C.

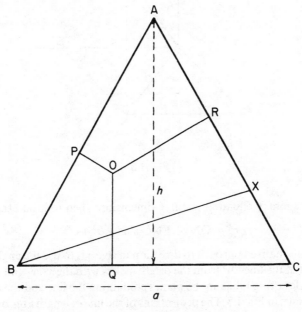

Fig. 1.2

1.7.3 *Three-dimensional geometry*

Many problems in science, and particularly in chemistry, involve more than two variables. These correspond to diagrams in three or more dimensions. They are often treated by taking two-dimensonal sections in planes perpendicular to each other. When lines or planes are perpendicular to each other they are called orthogonal.

In Fig. 1.3 the point P is distance r from origin O. The coordinates of P are (a, b, c), these being the projections of the line OP on to the x-, y- and z-axes. This is conveniently shown by drawing perpendicular, or normal, lines onto the z-axis and into the (x, y) plane. The projection ON in the (x, y) plane then gives projections OA $= a$ and OB $= b$ on the x- and y-axes.

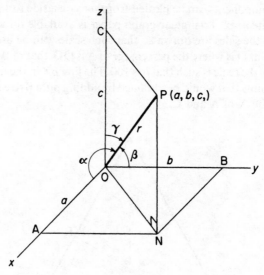

Fig. 1.3

Pythagoras's equation in three dimensions then follows because

$$r^2 = ON^2 + PN^2 = a^2 + b^2 + c^2.$$

An alternative approach to direction in space is to use the angles α, β and γ that the line OP from the origin makes with the x-, y- and z-axes, the angles being taken in the planes containing the line and the axis as also shown in Fig. 1.3. The projections of the line of length r onto the x-, y- and z-axes are then $a = r \cos \alpha, b = r \cos \beta$ and $c = r \cos \gamma$; the factors $\cos \alpha$, $\cos \beta$ and $\cos \gamma$ are called the direction cosines of OP and are denoted by l, m and n. Then by Pythagoras's equation, $l^2 + m^2 + n^2 = 1$.

An equation containing two variables represents a line or curve in two dimensions, and if it contains only first powers of the variables and no cross-terms it can be written in the form of equation (1.12) and represents a straight line. Similarly an equation in three variables represents a surface in three-dimensional space, and if it contains only the first powers of the variables and no cross-terms it represents a plane. The general equation of a plane is

$$\frac{x}{a} + \frac{y}{b} + \frac{z}{c} = 1, \tag{1.13}$$

as may be most easily checked by noting that the partial derivatives are then constants (Chapter 3). Equation (1.13) is that of a plane through the points A, B and C in Fig. 1.3, as can be seen by putting $y = z = 0$ to show that the intercept on the x-axis is $x = a$, and so on.

The direction of a plane in space is often required in crystallography, when it is denoted by the Miller indices (h, k, l). These are the smallest whole-number ratios of the reciprocals of the intercepts on the axes. Thus the plane through A, B and C in Fig. 1.3 is denoted by $(1/a, 1/b, 1/c)$, multiplied by a common factor if necessary so as to give the smallest whole numbers. The use of reciprocals here avoids infinite values because a plane parallel to an axis is defined to have a corresponding Miller index of zero.

1.7.4 Circle, ellipse, parabola and hyperbola

An equation in two variables, say x and y, which contains terms of the second degree (x^2, xy or y^2) but not higher, represents a circle, ellipse, parabola or hyperbola. The particular type of curve is readily identified from the relative values of the numerical coefficients.

Writing the general form of second-degree equation as

$$ax^2 + 2hxy + by^2 + 2gx + 2fy + c = 0, \tag{1.14}$$

the conditions for the various kinds of curve are

(a) Circle $a = b$ and $h = 0$,

(b) Ellipse $ab > h^2$,

(c) Parabola $ab = h^2$,

(d) Hyperbola $ab < h^2$.

At least one of the second-degree terms x^2, xy or y^2 must be present, but any other coefficients may be zero.

Example 1.6
A sample of perfect gas at fixed temperature obeys the relation $pV = a$ positive constant. Substituting p for y and V for x we have $a = b = 0$ and h is positive, so that $h^2 > ab$ and the curve obtained is a hyperbola (curve E in Fig. 1.4).

Example 1.7
The potential energy V for simple harmonic motion (Section 6.3) is given by $V = \frac{1}{2}kx^2$. Substituting V for y gives $a = k/2$ and $b = h(= 0)$, so that $ab = h^2 (= 0)$ and the curve is a parabola (curve F in Fig. 1.4).

There are standard forms of equation for each kind of curve. These are

(a) Circle $x^2 + y^2 = r^2$,   (1.15)

(b) Ellipse $x^2/\alpha^2 + y^2/\beta^2 = 1$,   (1.16)

(c) Parabola $y^2 = 4\alpha x$,   (1.17)

(d) Hyperbola $x^2/\alpha^2 - y^2/\beta^2 = 1$.   (1.18)

Example 1.8
Show that these standard forms satisfy the general conditions stated above.

Having identified the type of curve represented by any particular equation, its position and orientation relative to the coordinate axes may be found by looking for symmetry and for points of intersection. From equation (1.15) when $x = 0$, $y = \pm r$ and when $y = 0$, $x = \pm r$, so that the equation represents a circle of radius r centred at the origin (Curve A in Fig. 1.4).

An ellipse has major and minor axes with two foci on the long axis. Equation (1.16) gives $x = \pm \alpha$ when $y = 0$ and $y = \pm \beta$ when $x = 0$, so that the axes of the ellipse are the x- and y-axes. When $\alpha > \beta$ the x-axis is the longer one and so contains the two foci (curve B in Fig. 1.4).

The parabola represented by equation (1.17) passes through the origin ($y = 0$ at $x = 0$) and is symmetrical about the x-axis ($y = \pm 2\sqrt{(\alpha x)}$). Thus $y^2 = 4\alpha x$ gives curve C in Fig. 1.4 whereas $x^2 = 4\alpha y$ would give curve F.

A hyperbola lies between two intersecting straight lines called asymptotes. Curve E in Fig. 1.4 ($xy =$ constant) has the x- and y-axes as asymptotes, these being identified by seeing that as x tends to zero y tends to infinity, and as y tends to zero x tends to infinity. When, as here,

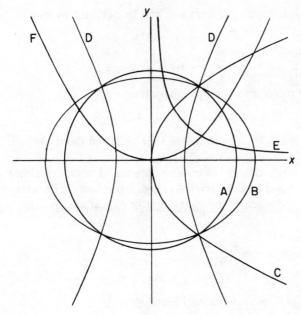

Fig. 1.4

the asymptotes are perpendicular to each other, the hyperbola is said to be rectangular. The general form, equation (1.18), cuts the x-axis at $\pm\alpha$ and does not cut the y-axis. It is symmetrical about the x-axis and has asymptotes given by $y/\beta = x/\alpha$ and $y/\beta = -x/\alpha$ (curve D in Fig. 1.4); these are perpendicular to each other if $\alpha = \beta$ and the hyperbola is then rectangular (see also Example 2.30).

Two operations may be necessary to convert any given equation into standard form: displacement of the origin and rotation of the axes. If the cross-product xy is present, axis rotation is needed to eliminate that term. The effect of this may be calculated analytically but can be seen by inspection in the simple cases usually met in physical applications.

Example 1.9
The equation
$$x^2 + y^2 - 2x + 4y = 4$$
represents a circle because $h = 0$ and $a = b$. The presence of terms in x and in y means that the centre of the circle is not at the origin. The equation is reduced to standard form by moving the origin to eliminate

these linear terms; the term in x can be obtained by using

$$(x-1)^2 = x^2 - 2x + 1$$

and that in y from

$$(y+2)^2 = y^2 + 4y + 4.$$

We can therefore write the equation as

$$(x-1)^2 + (y+2)^2 = 9$$

so that it is a circle with centre at $(1, -2)$ and radius 3.

The standard forms of equation are useful when experimental data are to be fitted to such curves. As remarked in Section 1.6, a linear graph is preferred when choosing the best fit to scattered points.

Example 1.10
Identify the curve

$$y = \alpha x(1-x)$$

and suggest two alternative linear forms.
    This equation has the form

$$\alpha x^2 - \alpha x + y = 0$$

so that $ab = h^2$ and the curve is a parabola. Differentiation (Section 2.8) shows the curve to have a maximum at $x = 1/2$, $y = \alpha/4$, and also $y = 0$ at both $x = 0$ and $x = 1$. The equation is reduced to standard form by writing

$$x^2 - x + y/\alpha = 0,$$

and using

$$(x-1/2)^2 = x^2 - x + 1/4$$

we obtain

$$(x-1/2)^2 = 1/4 - y/\alpha$$
$$= -(y-\alpha/4)/\alpha.$$

This becomes standard form by interchanging the variables and moving the origin to $(1/2, \alpha/4)$.
    A linear plot may be obtained either by plotting y against $x(1-x)$, which is obvious by inspection, or from y against $(x-1/2)^2$.
    Such a relation is often a good approximation to the properties of mixtures, such as the heat and volume changes on mixing liquids or on mixing gases. Departure from symmetry about $x = 1/2$ can be allowed for by using a series of terms in powers of $(x-1/2)$. A similar approach

is used to allow for anharmonicity of molecular vibrations in spectroscopy.

Example 1.11

Show that, neglecting air resistance and the curvature of the earth, a projectile moves in a parabola under the action of gravity.

Given velocity of projection u at an angle α to the horizontal, the vertical component of the initial velocity is $u \sin \alpha$. The vertical height y at time t is given by

$$y = ut \sin \alpha - \tfrac{1}{2}gt^2.$$

Simultaneously, the horizontal velocity is $u \cos \alpha$ so that the distance travelled in the horizontal direction is $x = ut \cos \alpha$. Eliminating t between these two equations gives

$$t = \frac{x}{u \cos \alpha},$$

$$y = x \tan \alpha - \frac{gx^2}{2u^2 \cos^2 \alpha},$$

so that $a = -g/(2u^2 \cos^2 \alpha)$ and $b = h = 0$. We therefore have $ab = h^2 \, (= 0)$ and the curve of y against x is a parabola.

1.7.5 Plane polar coordinates

Coordinates such as x and y, which are plotted as lengths on graphical scales, are called cartesian coordinates, in contrast to r and θ, a length and an angle, which are called polar coordinates. When the x- and y-axes are perpendicular to each other, they are called rectangular cartesian axes.

Point P in Fig. 1.5 may be defined either by (x, y) or by (r, θ), the relations between the coordinates being

$$x = r \cos \theta,$$

$$y = r \sin \theta. \tag{1.19}$$

Problems that possess circular symmetry will be solved most conveniently by using polar coordinates. The radius r is defined always to be positive so that x and y assume their proper signs in the various quadrants due to the signs of $\cos \theta$ and $\sin \theta$. The angle θ is generated by rotating the radius vector r from the x-axis in an anticlockwise direction; positive rotation is therefore conventionally anticlockwise.

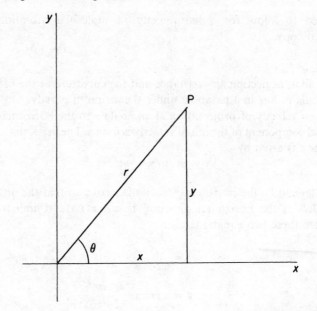

Fig. 1.5

1.8   Factorials and gamma functions

The factorial of a positive integer n is here denoted by $n!$ and is the product

$$n! = n(n-1)(n-2)(n-3)\ldots 1. \tag{1.20}$$

Thus

$$\frac{6!}{3!} = \frac{6 \times 5 \times 4 \times 3 \times 2 \times 1}{3 \times 2 \times 1} = 120.$$

It is conventional to define $0! = 1$ so that the general relation

$$(n+1)! = (n+1)(n!)$$

is still true when $n = 0$.

Since a graph could be plotted of $n!$ against n and a smooth curve drawn through the points for integral values of n, we wish also to be able to find the factorial of non-integral numbers. We anticipate subsequent chapters on integration to mention that this is achieved by use of the gamma function $\Gamma(x)$ defined in Section 6.4 as

$$\Gamma(x) = \int_0^\infty t^{x-1} e^{-t} dt \qquad (x > 0), \tag{1.21}$$

from which

$$x! = \Gamma(x + 1) = x\Gamma(x). \qquad (1.22)$$

Tables of this function are available, values being necessary only in the range $1 < x < 2$ since when $x > 2$ we use equation (1.22), thus

$$2.721! = \Gamma(3.721)$$

$$= 2.721\,\Gamma(2.721)$$

$$= (2.721)(1.721)\,\Gamma(1.721)$$

$$= 4.274 \qquad \text{from the tables.}$$

When n is very large, being typically a number of molecules in a sample of substance, an approximation that is often used is Stirling's theorem for $\ln n!$ in the form

$$\ln n! \simeq n \ln n - n, \qquad \text{for large } n. \qquad (1.23)$$

1.9 Probability

We express the probability of an event as a fraction between 0 and 1, where zero means that the event will not happen and unity means that the event is certain. A fundamental principle is that the probability of an event is proportional to the number of ways in which that event can occur. Thus in throwing two dice simultaneously we may obtain a total of 5 in four ways; by scoring any of (4, 1), (1, 4), (2, 3) or (3, 2). Each of the dice may fall in any one of six ways, and for any particular score on one we have six possibilities for the other; the total number of possible results is therefore found by multiplying 6 by 6 to give 36. The probability of scoring 5 is therefore $4/36 = 1/9$.

This argument applies to independent events, which are common in scientific applications. An important quantity is the number of ways of arranging N objects in line. For this we may choose the first as any one of the objects, and so in N ways assuming them to be distinguishable. The second is chosen from the remaining $N - 1$, and so in $N - 1$ ways. Since each choice of the first may be followed by any one of the choices of the second we multiply $N(N - 1)$ to find the number of ways of choosing the first two in the line. By repeating the argument we see that the required number is $N!$.

This is then extended to find the number of ways of placing N objects into boxes that contain $n_1, n_2, \ldots$ in successive boxes. We begin by placing the objects in line in $N!$ ways, and then divide the line into groups of $n_1, n_2, \ldots$. The objects in a group of n_1 may be rearranged in n_1! ways and these have so far been included in the total of N!. Since these rearrangements are not to be distinguished, we must remove them from the product by dividing by n_1!, hence the required number is $N!/(n_1! n_2! \ldots)$. If the objects are placed at random into the boxes, the probability of obtaining particular numbers in the boxes is then proportional to this factor; this is used in Example 2.17.

Example 1.12
The equilibrium constant for the reaction

$$H_2 + D_2 = 2HD$$

may be calculated from simple probability theory. We suppose that a catalyst surface, on which dissociation into atoms occurs, is exposed to a gas mixture containing equal numbers of H and D atoms. We assume that molecules are formed by random pairing of atoms on desorption from the surface. We then have equal probability of any particular atom being H or D. We may form H_2 (or D_2) only by both atoms being the same, but HD can be formed in two ways; either one site on the catalyst may be occupied by H and the other by D, or the other way round. We therefore form twice as many HD molecules as H_2 or D_2, giving for the equilibrium constant

$$K = \frac{[HD]^2}{[H_2][D_2]} = \frac{2^2}{1 \times 1} = 4,$$

which is a good approximation to the actual value of K at a high enough temperature.

1.10 Complex numbers

When equations are solved by simple algebraic methods we meet cases where no solution exists, that is to say no real solution. Thus the equation

$$x^2 + 4 = 0, \qquad x = \pm \sqrt{(-4)}$$

has no real roots. We can define i as the imaginary square root of -1 and write the solution as $x = \pm 2i$, which is the product of the real number ± 2 and the imaginary number i. This can be extended to include numbers having both real and imaginary parts, thus

$$x^2 - 4x + 8 = 0, \qquad x = 2 \pm \tfrac{1}{2}\sqrt{(-16)} = 2 \pm 2\sqrt{(-1)},$$

which is written as the complex numbers $2 \pm 2i$.

The definition and use of complex numbers in this way gives an advantage in the sense that an equation of degree n then always has n roots, some of which may be imaginary. These imaginary roots do not, however, have any physical meaning in the real world; complex-number theory cannot provide the physical scientist with any new concepts, but it does permit neater solutions to some problems. If, say, we are concerned with finding a boundary beyond which real solutions no longer exist, complex numbers can enable us to approach that boundary from the imaginary side as well as from the real one, and this has applications in stability problems. Complex numbers are valuable, moreover, as a general analytical tool.

In the general form of complex number $a + ib$ both a and b are real numbers; a is called the real part and ib the imaginary part. This should be regarded as an ordered pair of real numbers which have no relation between them. We must consider real and imaginary parts separately. Thus

$$(a + ib) - (c + id) = (a - c) + i(b - d). \tag{1.24}$$

When we have an equation relating complex numbers this really constitutes two simultaneous equations, because we can then equate the real parts and, separately, equate the imaginary parts. Thus

$$\text{if} \quad a + ib = c + id \quad \text{then} \quad a = b \quad \text{and} \quad c = d.$$

The imaginary number i follows the normal rules of algebra with the additional rule that $i^2 = -1$. Thus

$$(a + ib)(a - ib) = a^2 + b^2. \tag{1.25}$$

This also illustrates another important point, that we can obtain real numbers from combinations of imaginary ones, again only real numbers having physical meaning. When the sign is reversed in the complex number $z = (a + ib)$ to give $z^* = (a - ib)$ we call z^* the complex conjugate of z, and zz^* is the real number given by (1.25). This is used in wave mechanics, where the wavefunction ψ may be complex but $\psi\psi^*$ is a real probability.

It is sometimes useful to give a geometrical interpretation to complex numbers. This is done by plotting a two-dimensional graph showing real parts as abscissa and imaginary parts as ordinate. This is called the Argand diagram and is shown in Fig. 1.6 where point A represents the complex number $a + ib$. The complex conjugate A^* is then the reflection of the point A in the real axis. The complex number A can be regarded as the directed line, or vector, OA, and the addition of complex numbers A and B can be shown as if we add vectors OA and AC to give

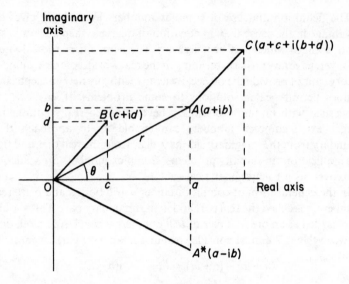

Fig. 1.6

OC. The length OA is called the modulus $|A|$ of the complex number A, given by

$$|A| = \sqrt{(a^2 + b^2)} \tag{1.26}$$

and this does not contain i and so is a real number.

When we choose to use polar coordinates we use the angle θ, called the argument of the complex number, and denote the modulus by r. We then have an alternative notation for complex numbers since $a = r \cos \theta$ and $b = r \sin \theta$, so that

$$a + ib = r(\cos \theta + i \sin \theta). \tag{1.27}$$

The analogy between complex numbers and vectors (Section 3.9) must not be pursued too far because other allowed operations with vectors do not apply to complex numbers. Thus multiplication of complex numbers gives

$$(a + ib)(c + id) = (ac - bd) + i(bc + ad) \tag{1.28}$$

or, in polar notation,

$$r_1(\cos \theta_1 + i \sin \theta_1) \times r_2(\cos \theta_2 + i \sin \theta_2)$$
$$= r_1 r_2 [\cos (\theta_1 + \theta_2) + i \sin (\theta_1 + \theta_2)]. \tag{1.29}$$

This means that on the Argand diagram we multiply by adding arguments $(\theta_1 + \theta_2)$ and multiplying the moduli $(r_1 r_2)$, so that multiplication and division involve rotating the vectors and changing their length.

There is a particularly useful relation between the exponential of a complex number and the trigonometric functions. By definition

$$e^x = 1 + x + \frac{x^2}{2!} + \frac{x^3}{3!} + \cdots,$$

and replacing the exponent by the complex number $x + iy$ gives

$$e^{x+iy} = e^x e^{iy}$$

and

$$e^{iy} = 1 + iy - \frac{y^2}{2!} - \frac{iy^3}{3!} + \cdots.$$

This series has alternating real and imaginary terms; the real terms are those in the series for $\cos y$ and the real coefficients of the imaginary terms are those for $\sin y$. Hence

$$e^{iy} = \cos y + i \sin y \tag{1.30}$$

and similarly

$$e^{-iy} = \cos y - i \sin y. \tag{1.31}$$

This gives us a third, exponential, form of notation for complex numbers. By comparing (1.30) and (1.31) with (1.27) we obtain

$$a + ib = r(\cos \theta + i \sin \theta) = r e^{i\theta},$$

$$r = \sqrt{(a^2 + b^2)}, \qquad \theta = \tan^{-1}(b/a). \tag{1.32}$$

This notation is valuable because exponential functions are particularly easy to manipulate. Thus the multiplication expressions (1.28) and (1.29) become

$$(a + ib)(c + id) = r_1 e^{i\theta_1} \times r_2 e^{i\theta_2} = r_1 r_2 e^{i(\theta_1 + \theta_2)}.$$

One of the principal advantages of complex numbers is that they enable trigonometric functions to be rewritten in terms of exponentials; trigonometric identities can then be readily established.

Example 1.13

$$\cos^2 y + \sin^2 y = (\cos y + i \sin y)(\cos y - i \sin y)$$

$$= e^{iy} e^{-iy} = e^0 = 1.$$

Addition or subtraction of (1.31) from (1.30) gives the useful substitutions

$$\cos x = \frac{e^{ix} + e^{-ix}}{2} \quad \text{and} \quad \sin x = \frac{e^{ix} - e^{-ix}}{2i}. \quad (1.33)$$

Example 1.14
Use equations (1.33) to show that $\cos^2 x - \sin^2 x = \cos 2x$.

The application to complex numbers of the calculus techniques described in the following chapters is straightforward, remembering to treat real and imaginary parts separately.

CHAPTER 2

Differential calculus

2.1 Significance and notation

Calculus refers repeatedly to behaviour in the limit of infinitesimal changes; a clear understanding of this concept, discussed in the next section, is essential. Calculus is valuable in science in general, and in chemistry in particular, for many reasons. Amongst these are that it enables us to study the effects of infinitesimal changes in conditions, effects which are often quite simple even though the effect of large finite changes may be quite complicated. In this way, problems can be separated into parts by examining separately the behaviour at a point from the behaviour over a wider range. Calculus also provides a clear and precise notation by which we can resolve problems that involve changes in many variables into ones involving changes in only one variable at a time. The techniques of differentiation and integration give connections between the values of properties and their rates of change, which are invaluable in the interpretation of experimental measurements. An appreciation of these uses of calculus requires not just the usual skills at the operations of differentiation and integration but an insight into the reasons why particular techniques are effective.

There is need to distinguish between the notation:

Δx, which means a large, finite or macroscopic change in x,

dx, which is the differential of x,

δx, which denotes an infinitesimal change in x, and

∂x which is the partial derivative sign.

We use the following functional notation:

$y = f(x)$ means that y is a function of x, or that given x, y is uniquely determined by an equation that contains both x and y but no other variables, only constants; another function of the same variables would be written as, for example, $g(x)$.

$y = f(u, v, w)$ means that y is a function of three variables u, v and w, any other terms being constants; it does not imply that u, v and w are independent.

$\dfrac{\mathrm{d}y}{\mathrm{d}x}$ and $y'(x)$ both denote the total derivative of y with respect to x.

$\dot{y}$ usually denotes differentiation of y with respect to time, or $\mathrm{d}y/\mathrm{d}t$.

$f'(x)$ is the total derivative of the function $f(x)$.

2.2 The calculus limit

The idea of a limit is fundamental to calculus and will be expressed both geometrically and analytically. In Fig. 2.1 a curve $y = f(x)$ is drawn with two points A and B on the curve separated by differences Δx in x and Δy in y. The line AB is a chord having a slope defined as $\Delta y/\Delta x$. When point B is moved down the curve to approach A, the slope of the chord will change until, when B coincides with A, we obtain the slope of the tangent to the curve at point A. The difference in x between points A and B is denoted by Δx and as B is moved to approach A this difference becomes infinitesimally small, being then denoted by δx; the slope of the chord becomes that of the tangent in the limit as $\delta x \to 0$.

We are quite deliberately considering what happens as $\delta x \to 0$, not at $\delta x = 0$; a way of interpreting that is to point out that when A and B are coincident we cannot obtain the tangent to the curve because the slope of a line through a single point can have any value. To find the slope of the tangent we need to take account of the shape of the curve on either side of the point. The limiting value of $\delta y/\delta x$ as $\delta x \to 0$ is called the derivative of y with respect to x, and is written as $\mathrm{d}y/\mathrm{d}x$, so that

$$\frac{\mathrm{d}y}{\mathrm{d}x} = \lim_{\delta x \to 0} \frac{\delta y}{\delta x}. \tag{2.1}$$

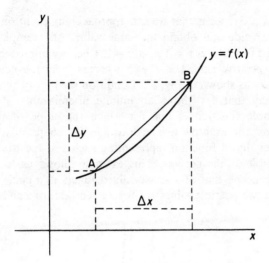

Fig. 2.1

The alternative analytical approach is to define δy as the change in y for change δx in x. Since $y = f(x)$ we can write

$$(y + \delta y) = f(x + \delta x)$$

and

$$\delta y = (y + \delta y) - y = f(x + \delta x) - f(x),$$

so that

$$\frac{dy}{dx} = \lim_{\delta x \to 0} \frac{f(x + \delta x) - f(x)}{\delta x}. \tag{2.2}$$

An alternative notation is to denote a difference in x by a new symbol such as h, and then allow h to tend to zero to give

$$\frac{dy}{dx} = \lim_{h \to 0} \frac{f(x + h) - f(x)}{h}.$$

The definition of the derivative given by (2.2) is used in the next section to obtain the rules for differentiation.

There are conditions that must be satisfied for calculus methods to be applicable to particular functions. In analytical terms the limit (2.2) must exist, the function then being said to be differentiable. A less severe condition concerns continuity; a function is said to be continuous if

$$\lim_{\delta x \to 0} f(x \pm \delta x) = f(x). \tag{2.3}$$

Condition (2.3) means that we can approach any point on the curve from either side and obtain the same value. An example of a discontinuous function is $y = 1/x$ as $x \to 0$; when we approach zero in x from the positive side, $1/x \to +\infty$ whereas from the negative side $1/x \to -\infty$, as shown in Fig. 2.2. Condition (2.3) is not then satisfied and we say that $1/x$ shows an infinite discontinuity at $x = 0$. A differentiable function is one for which the slope of the line is continuous; for example a line drawn in the shape of a square is continuous, but a function representing such a shape would not be differentiable at the corners. This example allows us to introduce another concept, that of a single-valued function; a circle or square could give two possible values of y for a given value of x and vice versa.

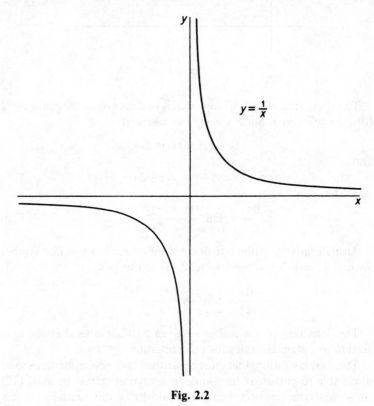

Fig. 2.2

Fig. 2.3 is drawn to illustrate other terms that are used to describe the behaviour of functions. The straight line $y = f(x) = x$ has the property

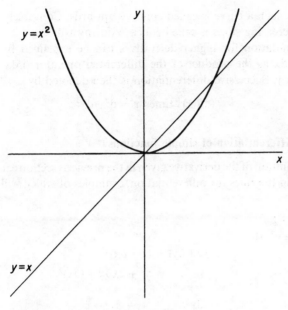

Fig. 2.3

that $f(-x) = -f(x)$, meaning that changing the sign of x changes the sign of y, and this is called an odd function of x. On the other hand, $y = x^2$ is a parabola (Section 1.7.4) for which $f(-x) = f(x)$; changing the sign of x does not change the sign of y and this is called an even function of x.

Most of the simpler functions used in physical applications are continuous, differentiable and single-valued, and the methods of calculus can then be used. A function such as $y = 1/x$ is 'well behaved' except as $x \to 0$, so that such functions can be used within certain ranges, which may have to be specified.

Having performed the operation of differentiation once we can do so again on the result of the first differentiation, assuming that also to be differentiable. This gives the second derivative of y with respect to x, which is written as d^2y/dx^2, and we can go on to higher derivatives $d^3y/dx^3, \ldots$. A straight line will have the same slope at any point on the line so that the first derivative is a constant; this means that the second derivative, which measures the rate of change of slope with change in x, is zero. The second derivative is thus seen to measure curvature. When the second derivative d^2y/dx^2 is positive, the slope is increasing with increase in x and the curve lies above its tangent as in

Fig. 2.1; such a curve is called concave upwards. Conversely, a curve with decreasing slope is called concave downwards.

The notation for higher derivatives can be explained by regarding dy/dx as the product of the differential operator d/dx and the variable y. Successive differentiation is then denoted by

$$(d/dx)^2 \text{ times } y = d^2y/dx^2.$$

2.3   Differentiation of simple functions

The definition of the derivative given in the previous section can be used to obtain the rules for differentiation, examples of which will now be given.

Example 2.1
Put $y = x^2$, then

$$(y + \delta y) = (x + \delta x)^2$$
$$= x^2 + 2x\delta x + (\delta x)^2.$$

Hence

$$\frac{dy}{dx} = \lim_{\delta x \to 0} \frac{(y + \delta y) - y}{\delta x}$$

$$= \lim_{\delta x \to 0} \frac{2x\delta x + (\delta x)^2}{\delta x}$$

$$= \lim_{\delta x \to 0} (2x + \delta x) = 2x$$

so that

$$\frac{dy}{dx} = \frac{d}{dx}(x^2) = 2x. \tag{2.4}$$

Example 2.2
Put $y = x^n$, then

$$(y + \delta y) = (x + \delta x)^n$$

$$= (x + \delta x)(x + \delta x)(x + \delta x) \ldots$$
$$= x^n + nx^{n-1}\delta x + \text{terms in } (\delta x)^2, \text{ etc.}$$

Hence

$$\frac{dy}{dx} = \lim_{\delta x \to 0} \frac{(y + \delta y) - y}{\delta x}$$

$$= \lim_{\delta x \to 0} (nx^{n-1} + \text{terms in } \delta x, \text{ etc.}).$$

When we take the limit all terms except the first contain δx and so tend to zero, giving

$$\frac{dy}{dx} = \frac{d}{dx}(x^n) = nx^{n-1}. \tag{2.5}$$

The formulae for differentiating a product or quotient may be derived from these first principles. If $y = uv$ where u and v are both functions of x

$$y + \delta y = (u + \delta u)(v + \delta v)$$

where δu and δv are the increments in u and v produced by an increment δx in x. Hence

$$\frac{dy}{dx} = \lim_{\delta x \to 0} \left(\frac{(u + \delta u)(v + \delta v) - uv}{\delta x} \right)$$

$$= \lim_{\delta x \to 0} \left(v \frac{\delta u}{\delta x} + u \frac{\delta v}{\delta x} + \frac{\delta u \delta v}{\delta x} \right).$$

Now the limit of a sum is the sum of the limits, and that of a product is the product of the limits. The first two terms contain the definitions of du/dx and dv/dx and the third gives

$$\left(\lim_{\delta x \to 0} \delta u \right) \left(\lim_{\delta x \to 0} \frac{\delta v}{\delta x} \right)$$

the first term of which is zero. Hence

$$\frac{dy}{dx} = \frac{d}{dx}(uv) = v \frac{du}{dx} + u \frac{dv}{dx}. \tag{2.6}$$

The quotient $y = u/v$ gives

$$\frac{dy}{dx} = \lim_{\delta x \to 0} \left[\left(\frac{u + \delta u}{v + \delta v} - \frac{u}{v} \right) / \delta x \right],$$

which may be written as

$$\frac{dy}{dx} = \lim_{\delta x \to 0} \frac{1}{v(v + \delta v)} \left(v \frac{(u + \delta u) - u}{\delta x} - u \frac{(v + \delta v) - v}{\delta x} \right)$$

so that

$$\frac{dy}{dx} = \frac{d}{dx}\left(\frac{u}{v}\right) = \frac{v \dfrac{du}{dx} - u \dfrac{dv}{dx}}{v^2}. \tag{2.7}$$

Example 2.3
Show that if

$$y = \frac{(1+x^2)}{x}$$

then

$$\frac{dy}{dx} = 1 - \frac{1}{x^2}.$$

2.4   The use of differentials; implicit differentiation

The derivative dy/dx can be interpreted not only as the product of the differential operator d/dx and the variable y, but also as the ratio of the differentials dy and dx. For this interpretation to be justified we must be able to apply the usual rules of algebra to differentials, and this we can in fact do. Introducing the notation $f'(x)$ for the result of applying the rules of differentiation derived in Section 2.3 to the function $f(x)$ we can write

$$\frac{dy}{dx} = f'(x)$$

and can then multiply both sides by dx to obtain

$$dy = f'(x)\,dx$$

and we call this the total differential of y. Notice that this is not called the differential with respect to any particular variable; the variable x appears only when we go on to write down the right-hand side of the equation, and indeed it can be changed if we so wish without affecting the left-hand side.

The rules for differentiation may then be written in the form

if	$y = x^n$	then	$dy = nx^{n-1}\,dx,$	(2.8)
if	$y = uv$	then	$dy = u\,dv + v\,du,$	(2.9)
if	$y = u/v$	then	$dy = (v\,du - u\,dv)/v^2.$	(2.10)

This notation is especially important when we cannot conveniently write y as an explicit function of x. Thus

$$y^2 - 2xy = 4$$

implies that y is a function of x, but an inconveniently complicated one. The derivative may be obtained most easily by differentiating each term to give

$$2y\,dy - 2x\,dy - 2y\,dx = 0$$
$$(y-x)dy = y\,dx$$
$$\frac{dy}{dx} = \frac{y}{y-x}.$$

When this method of obtaining derivatives is used the result will often contain both x and y, whereas when we can begin by obtaining y as an explicit function of x the result will be a function of x only. The original equation can be used to show that the results are the same.

Example 2.4
Show that implicit differentiation of $xy - x^2 = 1$ gives the same result as Example 2.3.
Differentiating the expression gives

$$x\,dy + y\,dx - 2x\,dx = 0$$
$$x\,dy = (2x - y)dx$$

$$\frac{dy}{dx} = \frac{2x-y}{x} = \frac{2x^2 - xy}{x^2} = \frac{x^2 + (x^2 - xy)}{x^2}.$$

But $x^2 - xy = -1$ so that, as before,

$$\frac{dy}{dx} = 1 - \frac{1}{x^2}.$$

Implicit differentiation may also be applied successively so as to obtain higher-order derivatives, given that differentiation of dx gives d^2x, of $x\,dx$ as a product gives $x\,d^2x + (dx)^2$ and, since $dx/dx = 1$, d^2x/dx^2 is the derivative of 1 and so is zero. From Example 2.4 we have $dy/dx = 1 - 1/x^2$ so that $d^2y/dx^2 = 2/x^3$. If we choose to use this as an example of successive implicit differentiation we also have

$$x\,dy + y\,dx - 2x\,dx = 0$$

and differentiating again gives

$$x\,d^2y + dx\,dy + dy\,dx + y\,d^2x - 2x\,d^2x - 2(dx)^2 = 0,$$

and division by $(dx)^2$ gives

$$x\frac{d^2y}{dx^2} + 2\frac{dy}{dx} + (y-2x)\frac{d^2x}{dx^2} - 2 = 0.$$

But $d^2x/dx^2 = 0$ so that

$$\frac{d^2y}{dx^2} = -\frac{2dy}{x\,dx} + \frac{2}{x}.$$

But $dy/dx = 1 - 1/x^2$ so that, as before,

$$\frac{d^2y}{dx^2} = \frac{2}{x^3}.$$

2.5   Logarithms and exponentials

The exponential function e^x, or $\exp(x)$, is defined by the series

$$e^x = 1 + x + \frac{x^2}{2!} + \frac{x^3}{3!} + \dots \qquad \text{for all } x. \qquad (2.11)$$

Differentiation of this series can be seen to give the same series as a result, so that when we differentiate e^x with respect to x we get e^x. Thus

$$\text{if} \quad y = e^x \quad \text{then} \quad dy = e^x\,dx, \quad \frac{dy}{dx} = e^x. \qquad (2.12)$$

The natural logarithm of y is defined by the relation

$$\text{if} \quad y = e^x \quad \text{then} \quad x = \ln y, \qquad (2.13)$$

and since e^x is positive (or zero) for all values of x, $\ln y$ is defined only for positive values of y. The notation $\ln$ for natural logarithm is used, with $\log_e$ as an alternative, to distinguish it from the Napierian or decadic logarithm $\log$ (or $\log_{10}$) which is defined by

$$\text{if} \quad y = 10^z \quad \text{then} \quad z = \log y. \qquad (2.14)$$

Since $\ln 10 = 2.3026$ (or $10 = e^{2.3026}$), the logarithms to the two bases are related by

$$y = 10^z = (e^{2.3026})^z = e^{2.3026z} \qquad (2.15)$$

so that

$$\ln y = 2.3026z = 2.3026 \log y. \qquad (2.16)$$

The derivative of a logarithm follows from the definition (2.14) and from the fact that the reciprocal of dy/dx is dx/dy. Thus

$$\text{if} \quad y = e^x, \quad \frac{dy}{dx} = e^x = y$$

$$\frac{dx}{dy} = \frac{1}{dy/dx} = \frac{1}{y},$$

which may also be written as

$$\text{if}\quad x = \ln y, \qquad dx = d(\ln y) = \frac{1}{y}dy, \qquad \frac{dy}{y} = d(\ln y). \quad (2.17)$$

Example 2.5
The rate of change of vapour pressure p of a liquid with temperature T is related to the heat of vaporization ΔH by

$$\frac{dp}{dT} = \frac{\Delta H}{RT^2}p.$$

This can be transformed by using both $dp/p = d(\ln p)$ and $d(1/T) = -(1/T^2)dT$ into

$$\frac{d(\ln p)}{d(1/T)} = -\frac{dp}{p}\frac{T^2}{dT} = -\frac{\Delta H}{R}.$$

Example 2.6
Show that if the rate coefficient k of a chemical reaction depends on temperature T according to

$$k = A\,e^{-E/RT}$$

where A and E are constants, then

$$\frac{d\ln k}{d(1/T)} = -\frac{E}{R}.$$

Example 2.7
Show that

$$\frac{d(\mu/T)}{d(1/T)} = \mu - T\frac{d\mu}{dT}.$$

2.6 The chain rule and differentiation by substitution

The chain rule is used when introducing a new dependent variable into a function that contains only one independent variable. Thus if y is known as a function of x, a two-dimensional graph can be drawn and dy/dx is defined. We may choose to introduce a new variable, say u, which is also a function of x. We then have three variables x, y and u but only one is independent and it is still a two-dimensional problem. If the new variable were not a function of only the existing ones, we would be moving into three dimensions, having then two independent variables, and would have to resort to partial differentiation as discussed in the next chapter.

Given the three variables x, y and u of which only one is independent, we can define three derivatives such as dy/dx, dy/du and du/dx (du/dy is simply the reciprocal of dy/du). These are related by the chain rule

$$\frac{dy}{dx} = \frac{dy}{du}\frac{du}{dx}. \tag{2.18}$$

This follows from the definition of a derivative as a limit and from the fact that the product of two limits is the limit of the product.

The introduction of a new dependent variable is called substitution, and its purpose in differentiation is to convert an otherwise complicated function into one to which the simple rules can be applied. This is best illustrated by examples.

Example 2.8
Let

$$y = (2x^2 - 1)^3.$$

The right-hand side could be evaluated algebraically and then differentiated term by term. A neater solution is to make the substitution

$$u = 2x^2 - 1,$$

then

$$du/dx = 4x$$

and

$$y = u^3, \qquad dy/du = 3u^2 = 3(2x - 1)^2.$$

Hence

$$\frac{dy}{dx} = \frac{dy}{du}\frac{du}{dx} = 12x(2x - 1)^2.$$

This can be expressed as a general formula, that

$$\text{if} \quad y = f(x)^n \quad \text{then} \quad dy = nf(x)^{n-1}f'(x)\,dx. \tag{2.19}$$

Example 2.9
Let $y = e^{ax}$, where a is a constant. Substitute

$$u = ax,$$

then

$$du/dx = a$$

and

$$y = e^u, \qquad dy/du = e^u.$$

Hence

$$\frac{dy}{dx} = \frac{dy}{du}\frac{du}{dx} = ae^u = ae^{ax}.$$

In general

$$\text{if} \quad y = e^{f(x)} \quad \text{then} \quad dy = e^{f(x)}f'(x)\,dx. \tag{2.20}$$

Example 2.10

Find the derivative with respect to temperature T of the expression

$$z = 1/(e^{hv/kT} - 1).$$

Substitute

$$u = (e^{hv/kT} - 1)$$

then

$$\frac{du}{dT} = -\frac{hv}{kT^2}e^{hv/kT}$$

and

$$z = \frac{1}{u}, \qquad \frac{dz}{du} = -\frac{1}{u^2} = -\frac{1}{(e^{hv/kT} - 1)^2}.$$

Hence

$$\frac{dz}{dT} = \frac{dz}{du}\frac{du}{dT} = \frac{hv}{kT^2}\frac{e^{hv/kT}}{(e^{hv/kT} - 1)^2}.$$

2.7   Turning points: maxima, minima and points of inflection

Fig. 2.4 is drawn to illustrate the various points A to E on a curve which can be detected by differential calculus and one point F which cannot.

Since the derivative dy/dx measures the slope of a curve, when $dy/dx = 0$ we have a horizontal tangent and y is said to be stationary. We then have a maximum as at A, a minimum as at B or a horizontal point of inflection as at C or D. These may be distinguished by successive differentiation.

The second derivative d^2y/dx^2 is the rate of change of slope. If this is negative the slope is decreasing as at A, where it is positive on the left and negative on the right, and point A is a maximum. Conversely, if d^2y/dx^2 is positive we have a minimum, as at B.

When both dy/dx and d^2y/dx^2 are zero we continue with successive

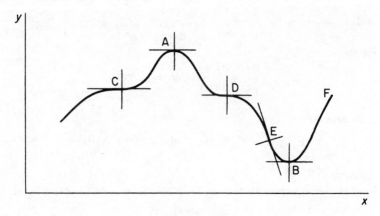

Fig. 2.4

differentiation. The conditions $d^2y/dx^2 = 0$ and d^3y/dx^3 positive are those for a minimum in the slope dy/dx. This corresponds to horizontal point of inflection C, where the slope decreases to zero and then increases again as we pass through the point. Point of inflection D is the converse, where the slope has a maximum value of zero being negative on either side, so that $dy/dx = d^2y/dx^2 = 0$ and d^3y/dx^3 is negative.

In general, we continue with successive differentiation until a non-zero derivative is found. When this is the nth derivative and n is even, the conditions for a maximum (negative derivative) or minimum (positive derivative) apply. When n is odd, the corresponding conditions for a horizontal point of inflection are used.

The conditions so far described are familiar ones, but points E and F also occur in physical chemistry. At E the slope is not zero but is nevertheless a minimum, as can be seen by noticing that from D onwards the slope becomes increasingly negative towards E and then increases (becomes less negative) beyond E. Point E is therefore characterized by dy/dx negative, d^2y/dx^2 zero and d^3y/dx^3 positive. The endpoint of an acid–base titration is at the corresponding point of inflection on a rising curve of pH against quantity, m say, of added base. These conditions are therefore $d(pH)/dm$ positive, $d^2(pH)/dm^2$ zero and $d^3(pH)/dm^3$ negative.

Point F is where the curve suddenly terminates; this is called an endpoint maximum. Since the definition of the derivative dy/dx requires that a limit is taken of the slope of a chord as the endpoints move closer together, such a limit is possible when approaching F from the left but not from the right. Such a one-sided derivative cannot be found analytically. This occurs when the vapour pressure of a liquid is plotted against temperature; the curve vanishes at the critical point.

Example 2.11
Given $y = x^4 + 4$,

$$\frac{dy}{dx} = 4x^3 = 0 \qquad \text{at } x = 0,$$

$$\frac{d^2 y}{dx^2} = 12x^2 = 0 \qquad \text{at } x = 0,$$

$$\frac{d^3 y}{dx^3} = 24x = 0 \qquad \text{at } x = 0,$$

$$\frac{d^4 y}{dx^4} = 24.$$

Hence the fourth derivative is the first one to be non-zero at the turning point $x = 0$. Since n is even and the derivative is positive the point $x = 0$, $y = 4$ is a minimum.

Example 2.12
Show that the curve $\ln y = x^3 + 1$ cuts the y-axis at a horizontal point of inflection.
    Differentiation of the equation gives

$$\frac{dy}{y} = 3x^2 \, dx,$$

$$\frac{dy}{dx} = 3x^2 y.$$

The second derivative may be obtained either by differentiating the equation a second time to give

$$-\frac{1}{y^2} dy^2 + \frac{1}{y} d^2 y = 6x \, dx^2 + 3x^2 \, d^2 x,$$

$$\frac{1}{y}\frac{d^2 y}{dx^2} = 6x + \frac{1}{y^2}\left(\frac{dy}{dx}\right)^2 = 6x + 9x^4$$

$$\frac{d^2 y}{dx^2} = 6xy + 9x^4 y,$$

or by differentiation of the expression for dy/dx to give, as before,

$$\frac{d^2y}{dx^2} = \frac{d}{dx}\left(\frac{dy}{dx}\right) = 6xy + 3x^2\frac{dy}{dx} = 6xy + 9x^4y,$$

and similarly

$$\frac{d^3y}{dx^3} = \frac{d}{dx}\left(\frac{d^2y}{dx^2}\right) = (6x + 9x^4)\frac{dy}{dx} + y(6 + 36x^3)$$

$$= 54x^3y + 27x^6y + 6y.$$

The point at which the curve cuts the y-axis is when $x = 0$, so that $\ln y = 1$, $y = e = 2.718$. From the above, when x is zero both dy/dx and d^2y/dx^2 are zero, but $d^3y/dx^3 = 6y = 16.3$ and so positive, and so we have a horizontal point of inflection of type C in Fig. 2.4.

Example 2.13

The velocity v of a chemical reaction is given by

$$v = k_1(a - x)^2 + k_2x(a - x),$$

where k_1, k_2 and a are positive constants, x is variable. Find the condition for a maximum in velocity, and the relation between k_1 and k_2 such that this occurs at $x = a/4$.

The required condition for a maximum is $dv/dx = 0$, d^2v/dx^2 negative, where

$$\frac{dv}{dx} = -2k_1(a - x) + k_2(a - 2x)$$

$$= 0 \qquad\qquad \text{if } 2k_1(a - x) = k_2(a - 2x),$$

$$\frac{d^2v}{dx^2} = 2(k_1 - k_2), \qquad \text{which is negative if } k_2 > k_1.$$

So that

$$\frac{k_1}{k_2} = \frac{a - 2x}{2(a - x)} \qquad \text{for maximum velocity.}$$

Put $x = a/4$, then

$$\frac{k_1}{k_2} = \frac{a/2}{3a/2} = \frac{1}{3} \qquad \text{for the maximum to be at } x = \frac{a}{4}.$$

This is the equation for a reaction which is second order in the reactant together with catalysis by-product.

Example 2.14
Find the ratio of height to diameter for a cylinder so that the surface area is a minimum for a given volume.

Put $V = \pi r^2 h$ and $A = 2\pi r h + 2\pi r^2$. We have two simultaneous equations to satisfy, that V is constant and A is a minimum. This may be solved in various ways. We could insert V as a constant and eliminate h by writing

$$h = V/\pi r^2$$

to give

$$A = 2V/r + 2\pi r^2$$

as an equation in the variables A and r, and use the condition $dA/dr = 0$. A neater solution is obtained by differentiating the expressions both for V and for A and putting $dV = 0$ for constant volume and $dA = 0$ for minimum surface area. Thus

$$dV = 2\pi r h\, dr + \pi r^2\, dh$$

$$= 0 \qquad \text{if } \frac{dr}{dh} = -\frac{r}{2h}$$

and

$$dA = 2\pi h\, dr + 2\pi r\, dh + 4\pi r\, dr$$

$$= 0 \qquad \text{if } \frac{dr}{dh} = -\frac{r}{h + 2r}$$

and these simultaneous equations give

$$2h = h + 2r, \qquad h = 2r$$

as the required condition. Thus the most economical shape to make a cylindrical tin is with the height equal to the diameter.

2.8 Maxima and minima subject to constraint; Lagrange's method of undetermined multipliers

Lagrange's method is a general one for solving problems in maxima and minima in the presence of constraints that can be expressed as simultaneous equations. A simple example is to find the point on a known curve that is nearest to the origin; the distance r will be a minimum when $dr = 0$, which, with the equation of the curve as a constraint, gives two simultaneous equations. The unknowns are the two coordinates of the required point so that the problem has a unique solution.

A related problem is to find not the maximum or minimum value of a given function, but the function itself such that a stationary condition applies. This is called the calculus of variations and is discussed in Section 5.7.

The procedure in Lagrange's method is to write each of the simultaneous equations in differential form, and to eliminate variables in the usual way by multiplying equations by factors and adding them together, the factors being the Lagrange multipliers. This provides an alternative method of solving problems in which we have as many simultaneous equations as variables, when numerical values can be obtained for the multipliers as in the following Examples 2.15 and 2.16. The method is particularly useful, however, when we have more variables than equations, a solution then being obtained in terms of multipliers whose values remain undetermined as in Example 2.17.

Example 2.15
Find the minimum distance from the origin to the curve $y = 2 - x^2$.

The first equation is obtained by expressing the distance r from the origin to any point (x, y) as

$$r^2 = x^2 + y^2$$

and minimizing by differentiating and putting $dr = 0$, to give

$$2r\,dr = 2x\,dx + 2y\,dy = 0$$

so that

$$x\,dx + y\,dy = 0. \tag{2.21}$$

The second equation is the constraint that the point (x, y) must lie on the curve. We express that also in differential form by writing the equation of the curve with zero on the right-hand side and differentiating. Thus

$$y - 2 + x^2 = 0,$$
$$dy + 2x\,dx = 0. \tag{2.22}$$

We solve these two simultaneous equations in the usual way by multiplying equation (2.22) by a constant factor α (the Lagrange multiplier) and adding to equation (2.21). This gives

$$x\,dx + y\,dy = 0$$

and

$$2\alpha x\,dx + \alpha\,dy = 0$$

so that

$$x(1 + 2\alpha)\,dx + (y + \alpha)\,dy = 0. \tag{2.23}$$

We now choose α so as to eliminate one of the differential variables, such as by putting

$$(1 + 2\alpha) = 0, \qquad \alpha = -1/2.$$

The other term in (2.23) must then also be zero, so that

$$(y + \alpha)\,dy = (y - \tfrac{1}{2})\,dy = 0$$

and, since dy may take any value, $y = \tfrac{1}{2}$. The equation of the curve then gives

$$x^2 = 2 - y = \tfrac{3}{2}$$

and the required distance is obtained from

$$r^2 = x^2 + y^2 = 7/4, \qquad r = \sqrt{7}/2.$$

Since in this case we have as many equations as unknowns, an algebraic solution of the simultaneous equations would give r as a function of x, so that

$$r^2 = x^2 + y^2 = x^2 + (2 - x^2)^2$$
$$r^2 = x^4 - 3x^2 + 4.$$

Differentiation then gives

$$2r\,dr = (4x^3 - 6x)\,dx$$

so that when $dr/dx = 0$, $4x^3 = 6x$, $x^2 = 3/2$, $r = \sqrt{7}/2$ as before.

Example 2.16
Find the shape of a right circular cone having minimum surface area for a given volume.

If the cone has base radius r, vertical height h and slant height a, we have $a^2 = r^2 + h^2$ and the volume V and surface area A are given by

$$V = \pi r^2 h/3$$
$$A = \pi r^2 + \pi r a.$$

For minimum A, $dA = 0$ giving

$$2\pi r\,dr + \pi r\,da + \pi a\,dr = 0,$$
$$(2r + a)\,dr + r\,da = 0. \qquad (2.24)$$

For a given volume, $dV = 0$ giving

$$\tfrac{2}{3}\pi r h\,dr + \tfrac{1}{3}\pi r^2\,dh = 0,$$
$$2h\,dr + r\,dh = 0. \qquad (2.25)$$

and from $r^2 + h^2 - a^2 = 0$

$$r\,dr + h\,dh - a\,da = 0. \tag{2.26}$$

We now take $(2.24) + \alpha(2.25) + \beta(2.26)$ to give

$$(2r + a + 2\alpha h + \beta r)\,dr + (r - \beta a)\,da + (\alpha r + \beta h)\,dh = 0.$$

Choosing α and β to eliminate the terms in da and dh, we put

$$r - \beta a = 0, \qquad \beta = r/a$$

and

$$\alpha r + \beta h = 0, \qquad \alpha = -\beta h/r = -h/a,$$

so that

$$2r + a - 2h^2/a + r^2/a = 0.$$

But $h^2 = a^2 - r^2$, so that

$$3r^2 + 2ra - a^2 = (3r - a)(r + a) = 0,$$

and since only positive values have physical significance, $a = 3r$.

An algebraic solution of the equations is possible but Lagrange's method gives the neater solution.

Example 2.17   The Boltzmann distribution
To determine the most probable distribution of molecules amongst quantum states: the Boltzmann distribution.

This is concerned with the most probable way in which a given amount of energy will be distributed amongst molecules. It combines the probability theory of Section 1.9 with maximization subject to constraints, and also uses Stirling's theorem, equation (1.23).

The physical model is to suppose that we have a large number of identical particles, each of which may be in any one of a large number of possible quantum states, each state having a characteristic energy. At any instant, we suppose a particular number n_i of particles to be in each quantum state i, and this is called a distribution. Probability enters into the argument by giving the number of ways of obtaining a particular distribution (particular numbers of particles in each state); it is this number of ways, or probability, that we seek to maximize subject to constraints.

This is an example of the use of Lagrange's method when the number of variables (the numbers $n_1, n_2, \ldots$) is very large. The argument turns on how many of these variables are independent, coupled with having an equation in which the sum of a series of independent terms is zero. In simplest form, if say $x + y = 0$ and if x and y are independent, then we must have $x = 0$ and $y = 0$; otherwise any non-zero value of x would require an equal and opposite value of y, but independence forbids that because the value of y is not determined by the value of x.

Suppose that we have a fixed total number N of particles, and each one has the same accessible set of quantum states, which have energies $\varepsilon_1, \varepsilon_2, \ldots, \varepsilon_i, \ldots$. We may regard the particles as objects and the quantum states as boxes; from Section 1.9 the number of ways of arranging N objects in boxes containing $n_1, n_2, \ldots, n_i, \ldots$ in each box is the probability P of that arrangement, given by

$$P = \frac{N!}{n_1! n_2! \ldots n_i! \ldots}. \tag{2.27}$$

The most probable arrangement is the one that can occur in the maximum number of ways; this is when $dP = 0$ or, more usefully, when

$$d \ln P = dP/P = 0$$

so that

$$d \ln P = d \ln N! - d \ln n_1! - d \ln n_2! - \ldots = 0. \tag{2.28}$$

If it were simply a matter of placing objects at random into boxes the most probable arrangement would be an equal number in each box. The problem is complicated by constraints, that the total number of particles is fixed and that the total energy is also fixed. These constraints give

$$N = n_1 + n_2 + \ldots + n_i + \ldots = \text{constant} \tag{2.29}$$

and since the n_i particles in the state with energy ε_i contribute $n_i \varepsilon_i$ to the total energy,

$$E = n_1 \varepsilon_1 + n_2 \varepsilon_2 + \ldots + n_i \varepsilon_i + \ldots = \text{constant}. \tag{2.30}$$

The variables in this problem are the numbers $n_1, n_2, \ldots, n_i, \ldots$. Since we have three simultaneous equations, all but two of these variables will be independent. We apply Lagrange's method by adding (2.28) to α times the differential form of (2.29) and β times the differential form of (2.30). The numbers $n_1, n_2, \ldots$ that we seek are average values, and so are treated as continuous variables rather than integers. Hence

$$dN = dn_1 + dn_2 + \ldots + dn_i + \ldots = 0 \tag{2.31}$$

and

$$dE = \varepsilon_1 \, dn_1 + \varepsilon_2 \, dn_2 + \ldots + \varepsilon_i \, dn_i + \ldots = 0. \tag{2.32}$$

Before doing the addition, we modify (2.28) by use of Stirling's approximation, equation (1.23)

$$\ln N! = N \ln N - N \qquad \text{for large } N,$$

giving

$$d \ln N! = \ln N \, dN.$$

Also, since N is constant, $dN = 0$, so that the term in N disappears from (2.28). This leaves

$$\ln n_1 \, dn_1 + \ln n_2 \, dn_2 + \ldots + \ln n_i \, dn_i + \ldots = 0. \qquad (2.33)$$

We now add $(2.33) + \alpha(2.31) + \beta(2.32)$ to obtain

$$(\ln n_1 + \alpha + \beta \varepsilon_1) \, dn_1 + (\ln n_2 + \alpha + \beta \varepsilon_2) \, dn_2 + \ldots$$
$$+ (\ln n_i + \alpha + \beta \varepsilon_i) \, dn_i + \ldots = 0. \qquad (2.34)$$

We now choose the multipliers α and β so as to eliminate n_1 and n_2. Equation (2.34) is then a sum of terms, each term containing only one of the independent variables n_i so that in order for the sum to be zero, each term must be zero. Hence

$$(\ln n_i + \alpha + \beta \varepsilon_i) \, dn_i = 0,$$
$$n_i = e^{-\alpha} e^{-\beta \varepsilon_i}.$$

The solution contains the undetermined multipliers α and β. The term $e^{-\alpha}$ may be written as a constant A, and physical reasoning identifies β with $1/kT$, giving the Boltzmann distribution law in the form

$$n_i = A e^{-\varepsilon_i / kT}.$$

2.9 Series

The use of series provides a valuable extension to algebraic methods. We can use a convergent series of terms to represent quantities that cannot be expressed in a simpler closed form, such as π, e or $\sin \theta$, and also problems that cannot be solved by simple algebra often yield to a solution in terms of series. The subject of series arises at this stage in our discussion because the series with most physical significance can be based upon Taylor's theorem, which is discussed in the next section.

A mathematical series contains terms that are related to each other, so that a formula exists for calculating any particular term. Such series are useful in practice only when successive terms become smaller and smaller so that the series converges. We may then use the series either to calculate its sum or as a source of successive numerical approximations.

2.9.1 Geometric series

This is a series of increasing powers of a variable in the form

$$s = 1 + x + x^2 + x^3 + \ldots \qquad (2.35)$$

Successive terms will become smaller and smaller, so that the series converges, if $-1 < x < 1$. The sum s of an infinite number of terms may be obtained by the device of multiplying each term in the series by x and subtracting the result from the original series, thus

$$sx = x + x^2 + x^3 + x^4 + \ldots \qquad (2.36)$$

Subtracting (2.36) from (2.35) we obtain

$$s(1-x) = 1, \qquad s = 1/(1-x).$$

Example 2.18

The theory of the heat capacity of a crystal gives rise to the series

$$1 + e^{-hv/kT} + e^{-2hv/kT} + e^{-3hv/kT} + \ldots$$

This can be seen to be a geometric series by making the substitution

$$x = e^{-hv/kT},$$

so that the sum of the series is

$$s = 1/(1 - e^{-hv/kT}).$$

2.9.2 Power series and Taylor's theorem

A power series is a general case of the geometric series discussed in the previous section, each of the successively higher powers of the variable being multiplied by a coefficient. In a mathematical power series the coefficients themselves form a series, and the convergence of the series depends on the ratio of successive coefficients being less than the reciprocal of the variable. In general terms a power series has the form

$$f(x) = a_0 + a_1 x + a_2 x^2 + a_3 x^3 + \ldots \qquad (2.37)$$

A convergent power series is one of the most useful ways of expressing many mathematical functions, and is often a useful and significant way of expressing the values of physical properties. If the series is stopped at x^n, it is then called a polynomial of degree n.

The values of the coefficients in a power-series expansion of a function can be expressed in terms of successively higher derivatives of the function. This is Taylor's theorem, which may be derived as follows:

Given a continuous function $f(x)$ whose value $f(a)$ is known at $x = a$ and which is successively differentiable; we denote the successive derivatives with respect to x by $f'(x), f''(x), \ldots$, and the values of these derivatives at $x = a$ by $f'(a), f''(a), \ldots$. We then assume that when x is increased from a to $a + h$ the increase in $f(x)$, which is $f(a + h) - f(a)$, can be expressed as a power series in h in the form

$$f(a + h) = f(a) + a_1 h + a_2 h^2 + a_3 h^3 + \ldots \qquad (2.38)$$

When x is expressed as $a + h$ the derivatives become

$$f'(x) = f'(a + h) = \frac{df(a + h)}{dx} = \frac{df(a + h)}{dh} \frac{dh}{dx},$$

but $x = a + h$ so that $dh/dx = 1$ and so

$$f'(x) = \frac{df(x)}{dh}.$$

The technique is then to differentiate (2.38) successively with respect to h, and then to put $h = 0$ to give expressions for the coefficients a_1, $a_2, \ldots$. Thus

$$f'(a + h) = \frac{d}{dh} (f(a) + a_1 h + a_2 h^2 + a_3 h^3 + \ldots)$$

$$= a_1 + 2a_2 h + 3a_3 h^2 + \ldots,$$

and when $h = 0$,

$$f'(a) = a_1.$$

Next

$$f''(a + h) = 2a_2 + 2 \times 3a_3 h + \ldots,$$

so that at $h = 0$

$$f''(a) = 2a_2.$$

Hence Taylor's theorem can be written in the form

$$f(a + h) = f(a) + hf'(a) + \frac{h^2}{2!} f''(a) + \frac{h^3}{3!} f'''(a) + \ldots \qquad (2.39)$$

When this series converges it provides successive approximations to the function $f(x)$ in terms of the values of the function and its derivatives at the point $x = a$.

This form of Taylor's theorem applies when we have only one independent variable x. With more than one independent variable the theorem is expressed in terms of partial derivatives, and this is discussed in Section 3.8.

A geometrical interpretation of Taylor's theorem is shown in Fig. 2.5. Point A is on the curve $y = f(x)$ at $x = a$. For increment h in x we obtain point B on the curve. The series (2.39) then provides successive approximations to an extrapolation of the curve from a known point A in terms of the known slope, curvature and successively higher derivatives, all at point A. The first approximation is

$$f(a + h) = f(a) + hf'(a), \qquad (2.40)$$

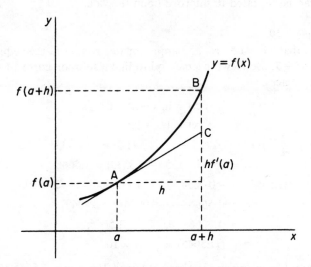

Fig. 2.5

which is point C on the tangent drawn from A, at $x = a + h$. This neglects any curvature, and so will be a poor approximation if curvature is present so that the slope $f'(x)$ changes with x. This change in slope is given by the second derivative and appears as the next term in the series. Higher derivatives allow for higher orders of curvature and give better approximations to the true shape of the curve.

Equation (2.40) is the basis of Newton's method of successive numerical approximation to a solution of the equation $f(x) = 0$. Given an approximate solution $x = x_0$, suppose $(x_0 + h)$ is the true root so that

$$f(x_0 + h) = 0.$$

We can find an approximate value of h by using the first approximation from Taylor's theorem that

$$f(x_0 + h) = f(x_0) + hf'(x_0)$$

so that

$$h = -\frac{f(x_0)}{f'(x_0)}$$

and our improved estimate of the root is

$$x_1 = x_0 + h = x_0 - \frac{f(x_0)}{f'(x_0)}. \tag{2.41}$$

This can be repeated to improve upon the value x_1.

Example 2.19
Given that $x = 1.5$ is an approximate root of the equation $\ln x = x^2 - 2$, use Newton's method to find a solution correct to four decimal places.
Put

$$f(x) = \ln x - x^2 + 2$$
$$f'(x) = 1/x - 2x.$$
$$f(1.5) = 0.1555, \qquad f'(1.5) = -2.3333$$
$$x_1 = 1.5 - 0.1555/(-2.3333) = 1.5666$$
$$f(1.5666) = -0.0054, \qquad f'(1.5666) = -2.4949,$$
$$x_2 = 1.5666 - (-0.0054)/(-2.4949)$$
$$= 1.5666 - 0.0022 = 1.5645,$$
$$f(1.5645) = -5.6 \times 10^{-6},$$

so that rapid convergence has been obtained. Hence $x = 1.5645$, correct to four decimal places.

2.9.3   Maclaurin's theorem

Maclaurin's theorem may be regarded as a special case of Taylor's theorem. Since Taylor's theorem gives the value of a function in terms of its value at a point and the derivatives at that point, and since the point at which the value of a function is often simplest is when x is zero, it is particularly useful to write the series for $a = 0$, so that $x = h$. This gives Maclaurin's theorem, which is

$$f(x) = f(0) + xf'(0) + \frac{x^2}{2!}f''(0) + \dots. \tag{2.42}$$

This may be used to develop series expansions of functions that can be successively differentiated. The following examples illustrate the expansion of particular functions.

Example 2.20

Let

$$f(x) = e^x, \qquad f(0) = e^0 = 1,$$

then

$$f'(x) = e^x, \qquad f'(0) = e^0 = 1,$$
$$f''(x) = e^x, \qquad f''(0) = 1,$$

so that the expansion becomes

$$e^x = 1 + x + \frac{x^2}{2!} + \frac{x^3}{3!} + \dots . \qquad (2.43)$$

This gives the useful approximation that, for small x, $e^x \simeq 1 + x$, the series being rapidly convergent for small values of x.

This is an example of the use of the theorem rather than a derivation of the series because e^x is defined as the series, as a result of which the derivative is e^x.

Example 2.21

Let

$$f(x) = \sin x \qquad f(0) = \sin 0 = 0,$$

then

$$f'(x) = \cos x, \qquad f'(0) = \cos 0 = 1,$$
$$f''(x) = -\sin x, \quad f''(0) = 0,$$
$$f'''(x) = -\cos x, \quad f'''(0) = -1,$$

so that

$$\sin x = x - \frac{x^3}{3!} + \frac{x^5}{5!} - \dots = x \quad \text{for small } x. \qquad (2.44)$$

Similarly we can obtain

$$\cos x = 1 - \frac{x^2}{2!} + \frac{x^4}{4!} - \dots . \qquad (2.45)$$

The hyperbolic functions $\sinh x$ and $\cosh x$ are defined by

$$\sinh x = (e^x - e^{-x})/2$$
$$= x + \frac{x^3}{3!} + \frac{x^5}{5!} + \dots$$
$$\cosh x = (e^x + e^{-x})/2$$
$$= 1 + \frac{x^2}{2!} + \frac{x^4}{4!} + \dots$$

and are useful because the terms do not alternate in sign as do those of $\sin x$ and $\cos x$.

These examples show that the expansions of the functions are consistent with the expressions for the derivatives. Maclaurin's theorem gives a new series, however, in the following example.

Example 2.22
Let

$$f(x) = \ln(1+x), \qquad f(0) = \ln 1 = 0,$$
$$f'(x) = 1/(1+x), \qquad f'(0) = 1,$$
$$f''(x) = -1/(1+x)^2, \qquad f''(0) = -1,$$

so that

$$y = \ln(1+x) = x - \frac{x^2}{2} + \frac{x^3}{3} - \frac{x^4}{4} + \dots \qquad (2.46)$$

This has been written as the variable y because the definition of the logarithm gives

$$y = \ln(1+x) \qquad \text{if} \qquad (1+x) = e^y$$

so that we may also write

$$e^y = (1+x) = 1 + y + \frac{y^2}{2!} + \frac{y^3}{3!} + \dots$$

so that

$$x = y + \frac{y^2}{2!} + \frac{y^3}{3!} + \dots \qquad (2.47)$$

Now equations (2.46) and (2.47) describe the same relation between x and y; they differ in that the first gives y as an explicit function of x, and the second gives x as an explicit function of y. This is an example of the inversion of a series, which is discussed further in the next section.

Example 2.23   The binomial series
This is a series expansion of $(1+x)^n$:

$$f(x) = (1+x)^n, \qquad f(0) = 1,$$
$$f'(x) = n(1+x)^{n-1}, \qquad f'(0) = n,$$
$$f''(x) = n(n-1)(1+x)^{n-2}, \qquad f''(0) = n(n-1).$$

Hence

$$(1+x)^n = 1 + nx + \frac{n(n-1)}{2!}x^2 + \frac{n(n-1)(n-2)}{3!}x^3 + \dots \qquad (2.48)$$

The successive coefficients of the powers of x are called binomial coefficients and have characteristic values, for example

$$(1+x)^4 = 1 + 4x + 6x^2 + 4x^3 + x^4.$$

The binomial series converges only when $-1 < x < 1$. When x is sufficiently small for x^2 not to be significant, we can neglect the third- and higher-order terms so that

$$(1+x)^n = 1 + nx \qquad \text{for small } x. \qquad (2.49)$$

Example 2.24
The binomial expansion gives a useful approximation when an expression contains an algebraic denominator, such as

$$\frac{1}{(a+x)^3} = \frac{1}{a^3(1+x/a)^3}$$

$$= \frac{1}{a^3}\left(1+\frac{x}{a}\right)^{-3}$$

$$= \frac{1}{a^3}\left[1 - \frac{3x}{a} + 6\left(\frac{x}{a}\right)^2 - 10\left(\frac{x}{a}\right)^3 + \dots\right],$$

which will be convergent so long as $x < a$, and terms beyond the second can be neglected if x/a is small enough compared with unity.

Example 2.25
Given the relation between the equilibrium constant K and the degree of dissociation α for a solution of a weak acid at concentration c

$$K = \frac{\alpha^2 c}{1-\alpha},$$

if we assume $(1-\alpha) \simeq 1$ we obtain $\alpha = \sqrt{(K/c)}$. Use the binomial expansion, assuming $K \ll c$, to show that successive approximations are

$$\alpha = \sqrt{(K/c)} - (K/c)/2 + (K/c)^{3/2}/8 - \dots.$$

Equation (2.49) is a quadratic equation for α in terms of K and c;

$$c\alpha^2 + K\alpha - K = 0,$$

so that the exact solution is given by formula (1.11) for solving a quadratic equation as

$$\alpha = -\frac{K}{2c} \pm \frac{1}{2c}\sqrt{(K^2 + 4Kc)}.$$

It is not possible by simple algebraic manipulation to separate out from this the term $\sqrt{(K/c)}$. We therefore resort to a series expansion by writing

$$\alpha = -\frac{K}{2c} \pm \sqrt{\left(\frac{K}{c}\right)}\left(1+\frac{K}{4c}\right)^{1/2},$$

so that the expression in the right-hand bracket is nearly unity, which means that the binomial expansion will converge rapidly. Hence

$$\alpha = -\frac{K}{2c} + \sqrt{\left(\frac{K}{c}\right)}\left(1+\frac{K}{8c}-\frac{1}{8}\frac{K^2}{16c^2}+\ldots\right)$$

$$\alpha = \sqrt{\left(\frac{K}{c}\right)}-\frac{K}{2c}+\frac{1}{8}\left(\frac{K}{c}\right)^{3/2}-\ldots$$

as required.

2.9.4 Inversion of a series

Given a convergent power series

$$y = x + a_2 x^2 + a_3 x^3 + \ldots \tag{2.50}$$

we can calculate y for a given value of x. Suppose, however, that we wish to find x for a given value of y. We could do so in particular cases by numerical trial and error, or we can invert (or revert) the series so as to obtain an expression for x as a function of y.

Assume that a power series exists for x in the form

$$x = y + b_2 y^2 + b_3 y^3 + \ldots$$

and substitute into equation (2.50)

$$y = (y + b_2 y^2 + b_3 y^3 + \ldots) + a_2(y + b_2 y^2 + b_3 y^3 + \ldots)^2$$
$$+ a_3(y + b_2 y^2 + b_3 y^3 + \ldots)^3 + \ldots. \tag{2.51}$$

For this to be true, the coefficients of the second and higher powers of y, which appear on expanding the right-hand side, must all be zero. Hence

$$b_2 + a_2 = 0, \qquad\qquad b_2 = -a_2,$$
$$b_3 + 2a_2 b_2 + a_3 = 0, \qquad b_3 = 2a_2^2 - a_3,$$

and so on. The series produced is

$$x = y - a_2 y^2 + (2a_2^2 - a_3)y^3 + (-5a_2^3 + 5a_2 a_3 - a_4)y^4$$
$$+ (14a_2^4 - 21a_2^2 a_3 + 6a_2 a_4 + 3a_3^2 - a_5)y^5 + \ldots. \tag{2.52}$$

The rate of convergence of the series (2.52) will not be the same as that of the original series (2.50) so that different numbers of terms will be necessary to obtain the same precision.

Example 2.26
Given the series (2.46), obtain the series (2.47).
By comparing (2.50) with (2.46) we obtain $a_2 = -1/2$, $a_3 = 1/3$, $a_4 = -1/4, \ldots$, so that (2.52) becomes

$$x = y + y^2/2 + y^3/6 + y^4/24 + y^5/120 + \ldots.$$

which is (2.47).

An alternative method of inverting a series is by successive approximations. We write equation (2.50) in the form

$$x = y - a_2 x^2 - a_3 x^3 - \ldots,$$

and substitute successive approximations for x into the right-hand side. The first approximation is $x = y$, which we substitute, retaining powers of y up to y^2, so that

$$x = y - a_2 y^2.$$

This is then the second approximation. We next substitute the second approximation and retain powers of y up to y^3

$$x = y - a_2 (y - a_2 y^2)^2 - a_3 y^3$$
$$= y - a_2 y^2 + (2a_2^2 - a_3) y^3,$$

and this is the third approximation. We continue by substituting the third approximation for x^2, the second approximation for x^3 and the first for x^4, retaining powers up to y^4, and so on. This method can be used for a convergent series of any kind.

Example 2.27
The equation of state of a real gas is a relation between pressure p, molar volume v and temperature T. The virial equation of state is a power series in the density $1/v$

$$pv = RT[1 + B(1/v) + C(1/v)^2 + D(1/v)^3 + \ldots].$$

An alternative is to use a power series in pressure in the form

$$pv = RT + B'p + C'p^2 + D'p^3 + \ldots,$$

and series inversion is used to obtain relations between the coefficients $B', C', D', \ldots$ and the virial coefficients $B, C, D, \ldots$.

Write the series in the form

$$p = RT(1/v) + B'p(1/v) + C'p^2(1/v) + D'p^3(1/v) + \dots$$

and the successive approximations are used to substitute for p on the right-hand side.

First approximation,

$$p = RT(1/v).$$

Substitute, retaining powers up to $(1/v)^2$.

Second approximation,

$$p = RT(1/v) + B'RT(1/v)^2.$$

Third approximation,

$$p = RT(1/v) + B'(1/v)(RT/v + B'RT/v^2) + C'(1/v)(RT/v)^2$$
$$= RT/v + B'RT/v^2 + (B'^2 + RTC')RT/v^3$$

Fourth approximation,

$$p = RT/v + B'(1/v)[RT/v + B'RT/v^2 + (B'^2 + RTC')RT/v^3]$$
$$\quad + C'(1/v)(RT/v + B'RT/v^2)^2 + D'(1/v)(RT/v)^3$$
$$= RT/v + RTB'/v^2 + (B'^2 + RTC')RT/v^3$$
$$\quad + [B'^3 + 3B'C'RT + D'(RT)^2]RT/v^4.$$

The required relations between the coefficients are

$$B = B',$$
$$C = B'^2 + RTC',$$
$$D = B'^3 + 3B'C'RT + D'(RT)^2,$$
$$\text{etc.}$$

2.9.5   *Empirical curve fitting by power series*

We may wish to fit a curve to experimental points so as to produce an economical expression for the data, or so as to interpolate and extrapolate in a way that is not subjective.

In the absence of any theoretical guidance as to the form of equation to be used, the normal practice is first to choose variables, and so scales for plotting the data, so as to obtain as near to a straight line as possible. A common practice is then to use a power series, the numerical values of the coefficients of which are found by curve fitting. This may be done in two ways. Given an equation

$$y = f(x) = a_0 + a_1 x + a_2 x^2 + \ldots ,$$

we can plot y against x and extrapolate to $x = 0$, the intercept being a_0. We can then calculate and plot $(y - a_0)/x$ against x, when the equation becomes

$$(y - a_0)/x = a_1 + a_2 x + a_3 x^2 + \ldots ,$$

so that the extrapolation to $x = 0$ gives a_1, and so on. This geometrical method has advantages in that the effect of successive terms is readily apparent and the increasing scatter of points which is inevitably produced shows when higher terms become irrelevant. The principal difficulty is that uncertain extrapolation is involved, the method depending on the estimation of limiting tangents.

The corresponding analytical procedure is to calculate values of the coefficients by the method of least squares, discussed in Section 7.8. In that case it is important to use the smallest possible number of terms. This is because a power series does not necessarily converge, and if more terms are used than are necessary the series can become oscillatory; a curve of complex shape that passes through every point merely reflects the accidental scatter of the data, a set of duplicate measurements producing a different curve.

2.10   The evaluation of limits by L'Hôpital's rule

We often need to find the limit of a function when the variables are taken either to zero or to infinity. This can usually be done by simple algebra, but difficulties arise when this leads to a product of zero and infinity, or to the ratios $0/0$ or ∞/∞, all of which are indeterminate.

Such difficulties may be resolved by the use of L'Hôpital's rule, which is that if two functions $f(x)$ and $g(x)$ are both zero at $x = a$ then

$$\lim_{x \to a} \frac{f(x)}{g(x)} = \lim_{x \to a} \frac{f'(x)}{g'(x)}. \tag{2.53}$$

This can be shown by using Taylor's theorem to expand the two functions

$$\frac{f(x)}{g(x)} = \frac{f(a) + (x-a)f'(a) + (x-a)^2 f''(a) + \ldots}{g(a) + (x-a)g'(a) + (x-a)^2 g''(a) + \ldots},$$

so that when $f(a) = g(a) = 0$ we may divide numerator and

denominator by $(x - a)$ to give

$$\frac{f(x)}{g(x)} = \frac{f'(a) + (x-a)f''(a) + \cdots}{g'(a) + (x-a)g''(a) + \cdots}$$

and as $x \to a$ the rule follows. It is most often used either when $a = 0$ or when $a = \infty$, and it applies equally well when the functions $f(x)$ and $g(x)$ both tend to infinity.

Example 2.28

$$\lim_{x \to 0} \frac{\sin x}{x}$$

would give 0/0 at $x = 0$, but differentiation of both the numerator and the denominator gives

$$\lim_{x \to 0} \frac{\cos x}{1} = 1.$$

Example 2.29
Evaluate

$$\lim_{x \to 0} (x \ln x).$$

This gives a product of zero and $-\infty$ at $x = 0$. We therefore write the expression as a quotient and apply L'Hôpital's rule:

$$\lim_{x \to 0} (x \ln x) = \lim_{x \to 0} \frac{\ln x}{1/x} = \lim_{x \to 0} \frac{1/x}{-1/x^2} = \lim_{x \to 0} (-x) = 0.$$

Example 2.30
Find the condition for the hyperbola

$$\frac{x^2}{a^2} - \frac{y^2}{b^2} = 1$$

to be rectangular (asymptotes perpendicular to each other).
    We have

$$b^2 x^2 - y^2 a^2 = a^2 b^2$$

and the asymptotes are straight lines of slope equal to the value of dy/dx as $x \to \infty$ and $y \to \infty$. By implicit differentiation

$$2b^2 x\,dx - 2a^2 y\,dy = 0,$$

$$\frac{dy}{dx} = \frac{b^2 x}{a^2 y},$$

which gives ∞/∞ as x and y tend to infinity. Differentiation of numerator and denominator with respect to x gives

$$\lim_{x \to \infty} \frac{dy}{dx} = \lim_{x \to \infty} \frac{b^2}{a^2 dy/dx}$$

so that in the limit,

$$(dy/dx)^2 = b^2/a^2, \qquad dy/dx = \pm b/a.$$

The asymptotes will be perpendicular if they have slopes of ± 1, so that the condition for a rectangular hyperbola is $b = a$.

Example 2.31
An alternative way of evaluating limits is by the use of series expansions. Example 2.10 gives an equation that occurs in the theory of the heat capacity of a monatomic solid. The high-temperature limit, as $T \to \infty$, of the expression

$$\frac{hv}{kT^2} \frac{e^{hv/kT}}{(e^{hv/kT} - 1)^2}$$

gives $0/0$. Expanding the exponentials in the form

$$e^{hv/kT} = 1 + hv/kT + (hv/kT)^2 + \cdots$$

shows that for very large T,

$$(e^{hv/kT} - 1) \to hv/kT \qquad \text{and} \qquad e^{hv/kT} \to 1,$$

so that the limit as $T \to \infty$ becomes k/hv.

2.11   The principles of Newtonian mechanics

Mechanics is concerned with force, mass, momentum and energy. The subject is deduced from a minimum number of first principles, the choice of which distinguishes one approach from another. In the Newtonian approach we regard force as fundamental, and this is usually the one adopted in introductory treatments because it can be applied directly to particular cases. An alternative is to regard energy as fundamental, as in the treatment of Hamilton and Lagrange, which is discussed in Section 4.7.

Force is a vector quantity, and the conditions for equilibrium under the action of forces is one application of the vector analysis discussed in Section 3.9.

Newtonian mechanics can be based on three principles: that when bodies interact with each other the forces of action and of reaction are equal in magnitude and opposite in direction; that a body changes its state of rest or uniform velocity only if acted upon by a force; and that

$$\text{force} = \text{mass} \times \text{acceleration}. \tag{2.54}$$

This will apply in any direction in space; if the velocity of a body in the x direction is $v = dx/dt$, a force F in the same direction will produce acceleration $dv/dt = d^2x/dt^2$ given by

$$F = m\frac{d^2x}{dt^2}. \tag{2.55}$$

Example 2.32

The conventional order of definition of mechanical properties is, first, velocity as rate of change of distance, then acceleration as rate of change of velocity and, finally, force as mass $\times$ acceleration. This order can be reversed with advantage in physical applications; we can use a transducer to measure the force that is exerted on a sufficiently massive body as a result of acceleration of a containing vehicle. Knowing force and mass we can calculate the acceleration; knowing acceleration as a function of time we can calculate velocity, and knowing velocity as a function of time we can calculate the distance travelled. This is the basis of inertial navigation systems.

In a time interval dt, the product $F\,dt$ will be

$$F\,dt = m\frac{d^2x}{dt^2}\,dt = m\frac{dv}{dt}\,dt = m\,dv = d(mv).$$

Anticipating our discussion of integration, this gives

$$\int_{t_1}^{t_2} F\,dt = \int_{mv_1}^{mv_2} d(mv) = mv_2 - mv_1.$$

The product mv is called momentum and the integral on the left-hand side is called impulse. If the force is constant the impulse becomes $F(t_2 - t_1)$, so that

$$\text{impulse} = \text{force} \times \text{time} = \text{change in momentum}. \tag{2.56}$$

We could equally well regard momentum as fundamental and obtain (2.54) as a consequence of (2.56).

When two or more bodies interact with each other the forces, and so the impulses, acting will be equal in magnitude and opposite in direction, so that the same applies to the changes in momentum. The total momentum will therefore be constant, which is the principle of conservation of momentum.

Energy may be defined as the result of doing work. When force F produces displacement dx in mass m,

$$F\,dx = m\frac{dv}{dt}\,dx = m\frac{dx}{dt}\,dv = mv\,dv.$$

Integration then gives

$$\int_{x_1}^{x_2} F\,dx = \int_{v_1}^{v_2} mv\,dv = \tfrac{1}{2}mv_2^2 - \tfrac{1}{2}mv_1^2.$$

The expression $\tfrac{1}{2}mv^2$ is called kinetic energy and the integral on the left is called work. If the force is constant the work done becomes $F(x_2 - x_1)$ so that

$$\text{work} = \text{force} \times \text{distance} = \text{change in kinetic energy.} \quad (2.57)$$

The forces with which we are concerned are often due to interaction with a field, as with mass in a gravitational field or electric charge in an electric field. The field strength is defined as the force on unit mass or charge. In the earth's gravitational field the force exerted on a body of mass m is called the weight w and the field strength is denoted by g, so that

$$w = mg.$$

By comparison with (2.54) we see that g can be identified as the acceleration of free fall under gravity.

The corresponding force F exerted by an electric field of strength E on a charge e is

$$F = Ee.$$

When a body is given a displacement dx in the direction opposite to the force F exerted on it by a field, work $F\,dx$ is done on the body and its energy increases. Energy that changes with position in a field is called potential energy and is denoted by V. Since force is in the direction of decreasing potential energy we have

$$dV = -F\,dx, \qquad F = -\frac{dV}{dx},$$

so that

$$\text{force} = -\text{potential gradient.} \quad (2.58)$$

If a body falls freely under gravity, or a charge is moved by an electric field, the decrease in potential energy is equal to the work done, which by (2.57) will be the increase in kinetic energy. Denoting kinetic energy by T and potential energy by V we have

$$T + V = \text{constant.} \tag{2.59}$$

For this to be true, all of the assumptions that have been made above must be true. In particular, the only effect of the action of force must be to increase the velocity of a body. This will be so only in the absence of friction. Movement through a resisting medium, such as a fluid, or sliding on a rough surface introduces a resisting force which reduces the acceleration of the body; some of the work done is converted into heat. At the molecular level, linear motion is converted into random thermal motion of the body and its surroundings, which is an irreversible process in the thermodynamic sense. Thus, although the conservation of momentum always applies to real processes, conservation of energy in the simple mechanical sense does not unless some formulation which includes heat is adopted.

The motion of a structured body in three-dimensional space, such as that of a molecule in a fluid, is resolved into linear motion (translation) of the centre of mass and rotation about the centre of mass. Rotational properties are expressed in terms of polar coordinates (Section 1.7.5). Our discussion of mechanics has, so far, neglected structure, as if the masses concerned were point particles. The arguments used apply to translation of the centre of mass of an extended body when it is treated as if all the mass were concentrated at that centre of mass, and all the forces were moved parallel to themselves so as to pass through that centre of mass. An applied force not passing through the centre of mass will tend to produce rotation of the body. We therefore introduce the principle of moments; the moment of a force about a point is defined as the product of the force and the perpendicular distance from the point to the line of action of the force. For an extended body to be in equilibrium under the action of forces, not only must the sum of the components of the forces in any direction in space be zero, but also the sum of the moments of the forces about any point must be zero, clockwise and anticlockwise moments being given opposite signs.

A molecule may be regarded as a set of discrete masses at the positions of the nuclei. Fig. 2.6 represents a diatomic molecule consisting of masses m_1 and m_2 rotating about an axis perpendicular to the paper through point O. Mass m_1 moves with tangential velocity v_1 in a circle of radius r_1 about O. In time interval dt the radius vector r_1

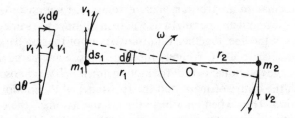

Fig. 2.6

moves through angle $d\theta$ and m_1 moves through an arc of length $ds_1 = r_1\,d\theta$. Then

$$v_1 = \frac{ds_1}{dt} = r_1\frac{d\theta}{dt} = r_1\omega, \qquad (2.60)$$

where ω is the angular velocity, $d\theta/dt$, measured in radian/second.

At constant angular velocity ω, the linear velocity v_1 will be constant in magnitude but changing in direction. From the triangle of velocities in Fig. 2.6, in time dt the mass m_1 gains velocity $v_1\,d\theta$ in the direction towards O; this corresponds to an acceleration towards the centre given by

$$\text{centripetal acceleration} = v_1\frac{d\theta}{dt} = v_1\omega = r_1\omega^2. \qquad (2.61)$$

If we suppose the masses to be connected by a weightless string, this must exert a force on m_1, or the mass exerts a force on the string given by

$$\text{centrifugal force} = m_1 r_1 \omega^2. \qquad (2.62)$$

If the string were elastic, or for a chemical bond, the bond length would increase with increase in ω, this being called centrifugal stretch. Since the string, or bond, must exert equal and opposite forces on the two masses we have

$$m_1 r_1 \omega^2 = m_2 r_2 \omega^2$$

and the position of the axis of rotation will be where

$$m_1 r_1 = m_2 r_2. \qquad (2.63)$$

This is the condition for finding centre of mass, so that a molecule that is freely rotating in space does so about its centre of mass.

A general motion of a rigid body in space, which includes both translation and rotation, can be expressed as the sum of translation of the centre of mass and rotation about the centre of mass. This may be seen by considering some particular line fixed in the body and perpendicular to an arbitrary axis of rotation; rotation through the same angle about any parallel axis will then produce the same angle of rotation of the line, together with translations of the line parallel and perpendicular to itself. We may therefore choose any parallel axis to describe the rotational part of the motion, for which we choose the one through the centre of mass, and the remainder of the motion will be translation. Thus when an athlete throws the hammer, the centre of mass of the hammer travels in a parabola (Example 1.11), the hammer meanwhile rotating about its centre of mass.

The kinetic energy T of our rotating diatomic molecule is the sum of the kinetic energies of the two masses, so that

$$T = \tfrac{1}{2}m_1 v_1^2 + \tfrac{1}{2}m_2 v_2^2$$
$$= \tfrac{1}{2}(m_1 r_1^2 + m_2 r_2^2)\omega^2.$$

The quantity in brackets is called the moment of inertia I of the molecule about an axis through the centre of mass and perpendicular to the axis of the molecule. Then

$$\text{kinetic energy of rotation} = \tfrac{1}{2}I\omega^2. \tag{2.64}$$

If the internuclear distance is $r = r_1 + r_2$ then (2.63) gives

$$r_1 = \frac{rm_2}{m_1 + m_2} \quad \text{and} \quad r_2 = \frac{rm_1}{m_1 + m_2}, \tag{2.65}$$

so that

$$I = \frac{m_1 m_2}{m_1 + m_2} r^2.$$

We define the reduced mass m^* by

$$\frac{1}{m^*} = \frac{1}{m_1} + \frac{1}{m_2}, \qquad m^* = \frac{m_1 m_2}{m_1 + m_2}, \tag{2.66}$$

so that

$$I = m^* r^2. \tag{2.67}$$

In the general case of a molecule composed of masses m_1, $m_2, \ldots$ held by chemical bonds at various positions in space, the centre of mass is a point such that for any axis through that point

$$\sum m_i r_i = 0,$$

where r_i is the perpendicular distance from m_i onto the axis. The moment of inertia about a particular axis is then

$$I = \sum m_i r_i^2.$$

When the molecule is planar we can choose rectangular coordinate axes OX, OY through the centre of mass as in Fig. 2.7. Mass m_1 at point P(x, y) will contribute $m_1 (PR)^2 = m_1 y^2$ to the moment of inertia about the x-axis OX. If we then choose arbitrary axes OX', OY' by rotating the axes through an angle θ, the point P will assume new coordinates (x', y') relative to OX', OY' given by

$$y' = PN = PM - MN = PM - QR = y\cos\theta - x\sin\theta,$$
$$x' = ON = OQ + QN = OQ + RM = x\cos\theta + y\sin\theta.$$

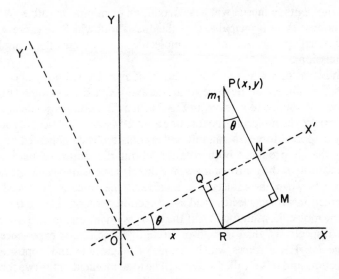

Fig. 2.7

The mass m_1 then contributes $m_1(y\cos\theta - x\sin\theta)^2$ to the moment of inertia about OX'. The total moment of inertia about the new axis OX' is then

$$I' = \sum m_i(y_i\cos\theta - x_i\sin\theta)^2.$$

The moment of inertia of our arbitrary planar molecule will therefore vary with the angle θ, and will have a stationary value if $dI'/d\theta = 0$. This is when

$$\frac{dI'}{d\theta} = -2\sum m_i(y_i\cos\theta - x_i\sin\theta)(y_i\sin\theta + x_i\cos\theta)$$
$$= -2\sum m_i x'_i y'_i = 0.$$

The expression $\sum m_i x_i y_i$ is called the product of inertia about the x- and

y-axes, and we see that if this is zero the moment of inertia will be a maximum or a minimum. The axes for which this applies are called the principal axes of the molecule, and the moments of inertia about the principal axes are called the principal moments of inertia of the molecule.

In three dimensions we find corresponding conditions, that when the axes are chosen so that the products of inertia about each pair of axes are zero,

$$\sum m_i x_i y_i = \sum m_i y_i z_i = \sum m_i z_i x_i = 0,$$

the moments of inertia will have maximum or minimum values. When we choose these principal axes as the ones about which we express the rotational properties of the molecule, we obtain the simplest expressions.

When a force acts to cause rotation of an extended body about an axis, the moment of the force about the axis is called the torque. Thus if force F is applied to mass m_1 in Fig. 2.6, in a direction perpendicular to the axis of the molecule, the torque about O is Fr_1. The effect of such a single applied force, as in a molecular collision, is complicated by the fact that it produces both rotation about the centre of mass and translation of the centre of mass. We therefore consider a simpler case, which is when the molecule is polar and the force is produced by interaction with an electric field in a direction perpendicular to the axis of the molecule, as in Fig. 2.8. If the charges on m_1 and m_2 are $+e$ and $-e$ respectively, then in field of strength E each mass experiences a force Ee. These forces will be equal in magnitude and opposite in direction, and this is called a couple. The total moment of the two forces about O is then

$$Eer_1 + Eer_2 = Ee(r_1 + r_2).$$

This moment depends only on the perpendicular distance $r = r_1 + r_2$ between the parallel forces, and so will be the same about any axis

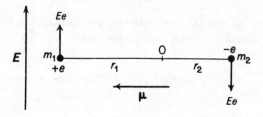

Fig. 2.8

perpendicular to the plane containing the forces. If the molecule were inclined at an angle θ to the direction of the field, as in Fig. 2.9, the perpendicular distance between the forces would be $r \sin \theta$ so that, in general,

$$\text{torque} = Eer \sin \theta = \mu E \sin \theta, \qquad (2.68)$$

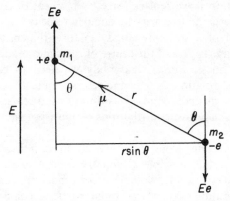

Fig. 2.9

the product $re = \mu$ being called the dipole moment of the molecule. Such a pair of equal and opposite forces will have no effect on the position in space of the centre of mass of the molecule because, when they are translated so as to pass through that centre, their resultant is zero. The effect of the field is therefore to apply torque and produce pure rotation of the molecule.

For linear motion equation (2.56) applies, so that

$$\text{force} = \text{rate of change of linear momentum.} \qquad (2.69)$$

In Fig. 2.8 the forces and the momenta are both vectors (Section 3.9) passing through the mass points and perpendicular to the axis of the molecule. We define angular momentum as the moment of the momentum, so that mass m_1 has linear momentum $m_1 v_1$ and angular momentum $m_1 v_1 r_1$ about an axis through O. From this definition and (2.69) we obtain

moment of force = torque
$$= \text{rate of change of angular momentum,} \qquad (2.70)$$
the total angular momentum of the two masses about O being

$$\text{angular momentum} = m_1 v_1 r_1 + m_2 v_2 r_2 = (m_1 r_1^2 + m_2 r_2^2)\omega = I\omega. \qquad (2.71)$$

The rotational kinetic energy of our molecule is simply the sum of the kinetic energies of the masses, so that

$$\text{kinetic energy} = \tfrac{1}{2}m_1 v_1^2 + \tfrac{1}{2}m_2 v_2^2 = \tfrac{1}{2}(m_1 r_1^2 + m_2 r_2^2)\omega^2 = \tfrac{1}{2}I\omega^2. \quad (2.72)$$

The relations used to describe angular motion are thus the same as for linear motion if we replace force by moment of force, mass by moment of inertia and velocity by angular velocity.

The motion of a structured body in three-dimensional space is resolved into translation of the centre of mass and motion relative to the centre of mass. For a rigid body we need to consider only translation and rotation, but for molecules we must include also the vibration of the chemical bonds. This vibration is discussed in Example 6.9, and makes additive contributions to both potential energy and kinetic energy.

Problems involving collisions between structureless particles are solved by using the principle of conservation of linear momentum. When rotation is possible, we have a corresponding conservation of angular momentum. When structured particles collide, equal and opposite forces act on each one at the point of contact. If we calculate the total angular momentum of the two bodies both about the same arbitrarily chosen axis, the forces acting on impact contribute equal and opposite moments about that axis. By equation (2.70) these produce equal and opposite rates of change of angular momentum, so that the total angular momentum about any axis remains unchanged.

In problems involving real molecules it may be necessary also to take account of the energies of allowed quantum states; these impose additional constraints, but the principles outlined above still apply.

CHAPTER 3

Differential calculus in three or more dimensions; partial differentiation

3.1 Significance and notation

Typical three-dimensional problems are the (p, v, T) properties of a gas or the vapour pressure of a two-component mixture. Here we have three variables and an equation connecting them, so that two variables are independent (see Section 1.4). In such cases the geometrical method is to draw a three-dimensional diagram and then to consider two-dimensional sections, which apply when the third variable is held constant. Thus we may consider p against v at constant T for a gas, or p against mole fraction x at constant T for the mixture. The corresponding analytical problem is: Given a function $z = z(x, y)$, how does z vary with x when y is kept constant, and with y when x is kept constant?

The general principle of taking two-dimensional sections of multi-dimensional space enables us to use the methods of differential calculus developed in the previous chapter. Given a relation $z = z(x, y)$, which defines a surface in three dimensions, constant y means a plane parallel to the (z, x) plane as in Fig. 3.1. If the surface cuts the plane in the curve shown, we can apply the methods of two-dimensional calculus to this curve. A tangent at point P will have a slope which is 'dz/dx in the plane at constant y'. This is written $(\partial z/\partial x)_y$, to be read as 'partial dz by dx at constant y'. The analytical definition of this partial derivative is:

If z is a continuous function of the independent variables x and y, then

$$(\partial z/\partial x)_y = \lim_{\delta x \to 0} \frac{z(x + \delta x, y) - z(x, y)}{\delta x} \tag{3.1}$$

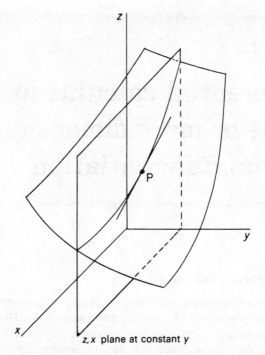

z, x plane at constant y

Fig. 3.1

where only the variable x is changed from x to $x + \delta x$, y being kept constant.

Analogously, we can define partial derivatives in the two other planes, at constant x and at constant z, denoted by $(\partial z / \partial y)_x$ and $(\partial y / \partial x)_z$. The derivatives in the previous chapter, where only two variables were concerned, are called total derivatives, and there are important differences between the properties of total and of partial derivatives. One property that they have in common is that the reciprocal of, say, $(\partial y / \partial x)_z$ is $(\partial x / \partial y)_z$ just as $1/(\mathrm{d}y/\mathrm{d}x) = \mathrm{d}x/\mathrm{d}y$.

When we have more than two independent variables, we define the partial derivative with respect to any one of the independent variables by keeping the others constant. Thus if z depends on x, y, u and v we can write

$$z = z(x, y, u, v)$$

and can define partial derivatives such as $(\partial z / \partial x)_{y, u, v}$, $(\partial z / \partial y)_{x, u, v}$, etc.

The rules for partial differentiation are the same as in the previous chapter, it being necessary only to remember that variables that are being held constant are treated in the same way as any other constants. For example, if

$$z = 2x^2 - 3xy + 7$$

then

$$(\partial z/\partial x)_y = 4x - 3y$$

and

$$(\partial z/\partial y)_x = -3x.$$

The method of implicit differentiation described in Section 2.4 can also be used. In the above example we can find $(\partial x/\partial y)_z$ by differentiating the equation at constant z, which gives

$$0 = 4x\,dx - 3x\,dy - 3y\,dx \qquad \text{at constant } z$$

so that

$$(4x - 3y)\,dx = 3x\,dy \qquad \text{at constant } z$$

$$(\partial x/\partial y)_z = 3x/(4x - 3y).$$

This example shows that the chain rule of Section 2.6 does not apply to the partial derivatives when a different variable is held constant in each case; with partial derivatives a minus sign appears, thus $(\partial z/\partial x)_y(\partial x/\partial y)_z = 3x = -(\partial z/\partial y)_x$. This relation is derived analytically in Section 3.6.

When a function is defined in terms of particular independent variables, such as $z(x, y)$, it will be clear that partial differentiation with respect to one variable implies that the other is to be kept constant. We can then use the abbreviated notation $\partial z/\partial x$ to imply constant y without ambiguity. In physical applications, however, we normally have a wide choice of independent variables. Thus for some property X, if we write $\partial X/\partial T$ it may mean at constant v or at constant p, these being different quantities (the relation between them is discussed in Section 3.6). We therefore need either to use the full notation, as $(\partial X/\partial T)_v$ and $(\partial X/\partial T)_p$, or to begin with an explanatory statement such as 'using independent variables v, T' so that $\partial X/\partial T$ then means $(\partial X/\partial T)_v$ and not $(\partial X/\partial T)_p$.

The higher-order differential coefficients such as $(\partial^2 z/\partial x^2)_y$ are a straightforward extension of the rules for total differentiation, but a new possibility also arises in the form of a mixed second derivative. This may be written in the form $\partial^2 z/\partial x\partial y$ or more explicitly as

$$\frac{\partial^2 z}{\partial x\partial y} \equiv \left(\frac{\partial}{\partial x}\left(\frac{\partial z}{\partial y}\right)_x\right)_y \equiv \left(\frac{\partial}{\partial x}\right)_y\left(\frac{\partial z}{\partial y}\right)_x.$$

These alternative forms all show that z is first to be differentiated with respect to y at constant x, the result then being differentiated with respect to x at constant y. From the above example we have

$$\frac{\partial z}{\partial y} = -3x, \qquad \frac{\partial^2 z}{\partial x \partial y} = -3.$$

If, on the other hand, the order of differentiation were reversed by differentiating first with respect to x at constant y and then with respect to y at constant x, the same result would be obtained. This illustrates an important principle, that under proper conditions the order of successive differentiation may be reversed. The properties of partial derivatives are so important in chemistry that we give alternative proofs of that principle, here and in the next section.

A geometrical interpretation of a mixed second derivative is that it measures the rate of change of slope as we move in a perpendicular direction over a surface. Using x and y as independent variables, the rate of change of slope $\partial z / \partial x$ with change in y is

$$\frac{\partial}{\partial y} \frac{\partial z}{\partial x}.$$

The derivative in reverse order is the rate of change of $\partial z / \partial y$ with change in x. We can move from a point $z(x, y)$ to a nearby point $z(x + \delta x, y + \delta y)$ by changing one variable at a time, corresponding to moving along two sides of a rectangle. Fig. 3.2 shows a projection of this into the (x, y) plane, the z-axis being perpendicular to the plane of

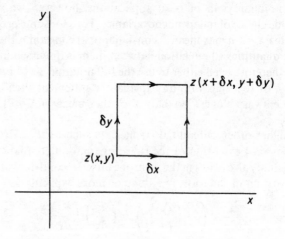

Fig. 3.2

the paper. By changing x first and then y the change in z is

$$\delta z = \frac{\partial z}{\partial x}\delta x + \left(\frac{\partial z}{\partial y} + \frac{\partial}{\partial x}\frac{\partial z}{\partial y}\delta x\right)\delta y,$$

whereas change first in y and then in x gives

$$\delta z = \frac{\partial z}{\partial y}\delta y + \left(\frac{\partial z}{\partial x} + \frac{\partial}{\partial y}\frac{\partial z}{\partial x}\delta y\right)\delta x.$$

In each case the expression in the brackets is the slope at $z(x, y)$ plus the increase in slope on moving by either δx or δy. The two expressions will be the same so long as the order of successive partial differentiation can be reversed. The expressions must be the same if a unique surface exists over which the alternative movements are made, and the conditions for that are that the variables shall be independent and the function and its derivatives shall be finite, continuous and single-valued.

3.2 An alternative approach to calculus

The interpretation of calculus as the behaviour of functions in the limit of infinitesimal changes that was adopted in the previous chapter is the more conventional one. An alternative approach, which has some advantages, is to consider the behaviour in terms of changes whose small magnitude is actually assessed at each stage of the argument. This regards the derivative not just in the limit but as differing from a ratio of infinitesimal quantities by a term that is zero to a number of decimal places. We write

$$\left(\frac{\partial z}{\partial x}\right)_y = \frac{z(x+h,y) - z(x,y)}{h} + 0(p) \tag{3.2}$$

where $0(p)$ means a term which is zero to p decimal places. As the increment h in x is made smaller, so p becomes larger. The advantage of this interpretation of a derivative is that we can then multiply both sides by h to obtain

$$h\left(\frac{\partial z}{\partial x}\right)_y = z(x+h,y) - z(x,y) + h0(p).$$

We are now able to write the mixed second derivative by applying the same rule; increase y by a small increment, for which we again choose h, and write the increase in the function again with an error $0(p)$. Hence

$$h^2\frac{\partial}{\partial y}\frac{\partial z}{\partial x} = h^2\frac{\partial^2 z}{\partial y\partial x} = [z(x+h,y+h) - z(x,y+h)]$$
$$- [z(x+h,y) - z(x,y)] + h^2 0(p). \tag{3.3}$$

When written in this form, the same result is obtained by partial differentiation in either order, because

$$h^2 \frac{\partial^2 z}{\partial x \partial y} = [z(x+h, y+h) - z(x+h, y)]$$

$$- [z(x, y+h) - z(x, y)] + h^2 0(p), \qquad (3.4)$$

and simple algebraic manipulation shows the right-hand sides of (3.3) and (3.4) to be the same. For this to apply, the variables x and y must be independent, so that we can increase either variable by h without change in the other variable.

3.3 The total differential

We have seen that if we have only two variables, and so two-dimensional diagrams, all derivatives are total derivatives, but with three or more variables we use partial derivatives. We have also seen that although the total derivative dy/dx denotes the operation of differentiation on the function $y = f(x)$, we can in fact separate this as the ratio of dy to dx. We shall now see how we can use a total differential of this kind even with more than two variables.

The geometrical interpretation of the total differential dz is that it is an increment in z, presupposing that it is such a small increment that any conclusions we draw are actually valid only in the usual calculus limit. We have seen in the previous section that we can if necessary place this on a firmer footing by stating explicitly that we are in error by a quantity that is zero to a certain number of decimal places.

The important thing about a total differential is that it is a real quantity. If we have an equation for a surface in three dimensions in the form $z = z(x, y)$, the meaning of dz is the increase in z when we move over the surface from one point on the surface to another one which is very close to it. It could be regarded as the increase in altitude when we walk obliquely across a hillside. This geometrical interpretation is here very useful because it is easily seen that we can move from one point to another on such a surface by a variety of routes, Consider a particular path which moves first in the plane at constant x, and then turns through a right-angle to move in the plane at constant y so as to reach the desired finishing point. The gradients in these two planes are given by the partial derivatives at constant x and at constant y respectively.

In two dimensions, the slope of a tangent is dy/dx, which is the increment in y divided by the increment in x. We then have the trivial relation

$$dy = \frac{dy}{dx} dx$$

meaning that the increment in y is the derivative times the increment in x. When the same relation is applied to a two-dimensional section of multidimensional space it becomes

$$dz = \left(\frac{\partial z}{\partial x}\right)_y dx.$$

It is important to notice that this expression contains a mixture of total and partial differentials. The total differentials dz and dx are potentially real increments; dz is, so far, the increase in z when we move distance dx in the plane at constant y.

Combining this expression with the principle that we can make a movement over a surface by movements in perpendicular planes, the geometrical argument shows that we then add the increments in z produced by the component moves. Thus

$$dz = \left(\frac{\partial z}{\partial x}\right)_y dx + \left(\frac{\partial z}{\partial y}\right)_x dy. \tag{3.5}$$

This is the definition of the total differential of $z = z(x, y)$. When we have more variables a term is included for each, so that if

$$z = z(x, y, u, v, \ldots)$$

then

$$dz = \left(\frac{\partial z}{\partial x}\right)_{y, u, v, \ldots} dx + \left(\frac{\partial z}{\partial y}\right)_{x, u, v, \ldots} dy + \ldots \tag{3.6}$$

Example 3.1
The pressure of a gas depends on the temperature T and molar volume v. Thus $p = p(T, v)$ so that

$$dp = \left(\frac{\partial p}{\partial T}\right)_v dT + \left(\frac{\partial p}{\partial v}\right)_T dv.$$

This equation means that the differential change in pressure produced by simultaneous differential changes dT in T and dv in v is given by the right-hand side of the equation. These differential changes do not have any physical significance because they need to be integrated (Chapter 4) before they represent real changes, and integration will be possible only when we know the physical properties of the gas. We cannot, of course, predict such properties by purely mathematical means.

If we make the physical assumption that the gas is perfect we then have $pv = RT$, in which case

$$\left(\frac{\partial p}{\partial T}\right)_v = \frac{R}{v} = \frac{p}{T} \quad \text{and} \quad \left(\frac{\partial p}{\partial v}\right)_T = -\frac{RT}{v^2} = -\frac{p}{v}$$

so that in this case

$$\mathrm{d}p = \frac{p}{T}\mathrm{d}T - \frac{p}{v}\mathrm{d}v \quad \text{or} \quad \frac{\mathrm{d}p}{p} = \frac{\mathrm{d}T}{T} - \frac{\mathrm{d}v}{v}.$$

The definition of a total differential is quite general, and applies whether or not the variables are independent. This will be shown in the next section since it provides useful practice in handling differentials. The definition of the total differential is the basis from which other useful relations are derived, so that we can use these relations even when we do not use only independent variables, which may happen in physical applications either by accident or by design.

3.4   General expression for a total differential

The geometrical justification of the definition of a total differential given in the previous section was based upon the assumption that a surface could be drawn in space of dimensions equal to the number of variables. The existence of an equation for the surface means that all but one of the variables can be independent. It was also stated that the definition applies even when the variables are not independent and this will now be shown.

Suppose a quantity z depends upon the independent variables x and y,

$$z = z(x, y), \qquad x \text{ and } y \text{ independent.} \tag{3.7}$$

Suppose we now choose to introduce another variable w and write

$$z = z(x, y, w). \tag{3.8}$$

We now have four variables, z, x, y and w, but still only two of them are independent. This means that a relation must exist between w, x and y, which we write as

$$w = w(x, y). \tag{3.9}$$

Since z may now be expressed as a function of any two of the variables we may also write

$$z = z(x, w) \tag{3.10}$$

and

$$z = z(y,w). \tag{3.11}$$

Since equations (3.7), (3.9), (3.10) and (3.11) all express w or z as a function of two variables, we can write the corresponding expressions for the total differential dz:

from (3.7)

$$dz = \left(\frac{\partial z}{\partial x}\right)_y dx + \left(\frac{\partial z}{\partial y}\right)_x dy, \tag{3.12}$$

from (3.9)

$$dw = \left(\frac{\partial w}{\partial x}\right)_y dx + \left(\frac{\partial w}{\partial y}\right)_x dy, \tag{3.13}$$

from (3.10)

$$dz = \left(\frac{\partial z}{\partial x}\right)_w dx + \left(\frac{\partial z}{\partial w}\right)_x dw, \tag{3.14}$$

from (3.11)

$$dz = \left(\frac{\partial z}{\partial y}\right)_w dy + \left(\frac{\partial z}{\partial w}\right)_y dw. \tag{3.15}$$

Each of these expressions contains the total differential of three of the four quantities z, x, y and w. In particular, (3.14) and (3.15) will remain true if we choose to hold the fourth quantity constant. Dividing (3.14) by dx at constant y we obtain

$$\left(\frac{\partial z}{\partial x}\right)_y = \left(\frac{\partial z}{\partial x}\right)_{w,y} + \left(\frac{\partial z}{\partial w}\right)_{x,y}\left(\frac{\partial w}{\partial x}\right)_y, \tag{3.16}$$

and dividing (3.15) by dy at constant x gives

$$\left(\frac{\partial z}{\partial y}\right)_x = \left(\frac{\partial z}{\partial y}\right)_{w,x} + \left(\frac{\partial z}{\partial w}\right)_{y,x}\left(\frac{\partial w}{\partial y}\right)_x. \tag{3.17}$$

Substitution of (3.16) and (3.17) into (3.12) and using (3.13) gives

$$dz = \left(\frac{\partial z}{\partial x}\right)_{w,y} dx + \left(\frac{\partial z}{\partial y}\right)_{w,x} dy + \left(\frac{\partial z}{\partial w}\right)_{y,x} dw, \tag{3.18}$$

which would be obtained if we had assumed (3.8) in the first place and written the total differential to include the superfluous variable.

The number of independent variables that properly apply to a particular application is a physical problem, not a mathematical one. The significance of the above is to show that we do not make any mathematical errors as a result of a physical error in choosing that number.

3.5   Exact differentials

This term applies to an expression of the form $X \, dx + Y \, dy$, and means that a quantity, z say, exists whose total differential is of that form, so that

$$dz = X \, dx + Y \, dy. \tag{3.19}$$

This will be so if the relation (3.19) is identical with

$$dz = \left(\frac{\partial z}{\partial x}\right)_y dx + \left(\frac{\partial z}{\partial y}\right)_x dy \tag{3.20}$$

and the test for exactness is to see if this is so. This can be done by taking the mixed second derivative and checking that the order of differentiation can be reversed. Thus to confirm that

$$X = \left(\frac{\partial z}{\partial x}\right)_y \qquad \text{and} \qquad Y = \left(\frac{\partial z}{\partial y}\right)_x$$

we calculate

$$\left(\frac{\partial X}{\partial y}\right)_x = \frac{\partial^2 z}{\partial y \, \partial x} = \left(\frac{\partial Y}{\partial x}\right)_y. \tag{3.21}$$

This operation is called cross-differentiation, and is true for exact differentials.

Example 3.2
Show that $(2xy \, dx + x^2 \, dy)$ is an exact differential, whereas $(x^2 \, dx + 2xy \, dy)$ is not.

Applying the cross-differentiation test, equation (3.21),

$$\frac{\partial}{\partial y}(2xy) = 2x$$

$$\frac{\partial}{\partial x}(x^2) = 2x$$

and the test succeeds with the first expression whereas from the second expression

$$\frac{\partial}{\partial y}(x^2) = 0$$

$$\frac{\partial}{\partial x}(2xy) = 2y$$

and the test fails.

Example 3.3

The volume v of a sample of gas depends on the temperature T and the pressure p. Thus $v = v(T, p)$ and

$$dv = \left(\frac{\partial v}{\partial T}\right)_p dT + \left(\frac{\partial v}{\partial p}\right)_T dp. \tag{3.22}$$

If the volume is changed by dv at constant pressure, the work done on the gas is $-p\,dv$. Substitution for dv from equation (3.22) gives

$$\text{work} = -p\,dv = -p\left(\frac{\partial v}{\partial T}\right)_p dT - p\left(\frac{\partial v}{\partial p}\right)_T dp.$$

We can show that work is not an exact differential by applying the cross-differentiation test, equation (3.21)

$$\left(\frac{\partial}{\partial p}\left[-p\left(\frac{\partial v}{\partial T}\right)_p\right]\right)_T = -\left(\frac{\partial v}{\partial T}\right)_p - p\frac{\partial^2 v}{\partial p\,\partial T},$$

$$\left(\frac{\partial}{\partial T}\left[-p\left(\frac{\partial v}{\partial p}\right)_T\right]\right)_p = -p\frac{\partial^2 v}{\partial p\,\partial T}.$$

These would be identical only if $(\partial v/\partial T)_p$ were zero. This is not physically possible because the coefficient of expansion $(1/v)\,(\partial v/\partial p)_T$ is always finite and positive.

The geometrical significance of this is that for an exact differential a unique surface exists when z is plotted against x and y. If we move from one point (x_1, y_1) on the surface to another (x_2, y_2) by changes in x and y, the change in z is uniquely defined by the initial and final values of x and y. Conversely, an inexact differential does not define a unique surface and the change in z will depend on the particular path followed from (x_1, y_1) to (x_2, y_2). A thermodynamic function of state is one that can be expressed as an exact differential, so that changes in the function are independent of path. The above example shows that work is not a thermodynamic function of state, which is very significant.

Example 3.4

Cross-differentiation of an exact differential is used to obtain the Maxwell relations in thermodynamics from the Gibbs equations. Given the exact differential

$$dG = v\,dp - S\,dT,$$

cross-differentiation gives

$$\left(\frac{\partial S}{\partial p}\right)_T = -\left(\frac{\partial v}{\partial T}\right)_p.$$

When we have more than two independent variables, a straightforward extension of the above argument gives, for $z = z(u, v, w, \dots)$

$$dz = \left(\frac{\partial z}{\partial u}\right)_{v, w, \dots} du + \left(\frac{\partial z}{\partial v}\right)_{u, w, \dots} dv + \left(\frac{\partial z}{\partial w}\right)_{u, v, \dots} dw + \dots$$

and the cross-differentiation test for an exact differential

$$A\, du + B\, dv + C\, dw + \dots$$

becomes

$$\left(\frac{\partial A}{\partial v}\right)_{u, w, \dots} = \left(\frac{\partial B}{\partial u}\right)_{v, w, \dots},$$

$$\left(\frac{\partial B}{\partial w}\right)_{u, v, \dots} = \left(\frac{\partial C}{\partial v}\right)_{u, w, \dots},$$

and so on, for any pair of terms.

3.6 Relations between partial derivatives

We can use the definition of a total differential given in Section 3.3 to obtain useful relations between partial derivatives. If $z = z(x, y)$ then

$$dz = \left(\frac{\partial z}{\partial x}\right)_y dx + \left(\frac{\partial z}{\partial y}\right)_x dy. \tag{3.23}$$

At constant z, $dz = 0$, so that

$$\left(\frac{\partial z}{\partial x}\right)_y dx = -\left(\frac{\partial z}{\partial y}\right)_x dy, \qquad \text{at constant } z.$$

Dividing by dy at constant z

$$\left(\frac{\partial z}{\partial x}\right)_y \left(\frac{\partial x}{\partial y}\right)_z = -\left(\frac{\partial z}{\partial y}\right)_x, \tag{3.24}$$

which is the chain rule for partial derivatives mentioned in Section 3.1.

Example 3.5

Equation (3.24) can be used to obtain purely mathematical relations between experimental properties that are defined as partial derivatives.

The coefficient of expansion $\alpha = (1/v)(\partial v/\partial T)_p$, isothermal compressibility $\kappa = -(1/v)(\partial v/\partial p)_T$ and thermal pressure coefficient $\gamma = (\partial p/\partial T)_v$ are related by

$$\kappa\gamma = -\frac{1}{v}\left(\frac{\partial v}{\partial p}\right)_T\left(\frac{\partial p}{\partial T}\right)_v = \frac{1}{v}\left(\frac{\partial v}{\partial T}\right)_p = \alpha,$$

so that any experimental measurements of these three quantities which did not obey this relation would certainly be wrong. The relation is useful also in that any one quantity can be obtained by measuring the other two.

In the same way, the heat capacity at constant pressure $C_p = (\partial H/\partial T)_p$, isenthalpic Joule–Thomson coefficient $\mu = (\partial T/\partial p)_H$ and isothermal Joule–Thomson coefficient $\phi = (\partial H/\partial p)_T$ are related by $\phi = -\mu C_p$.

If a relation exists between the variables x and y in equation (3.23), only one variable is independent and we can write the total differential of z with respect to either one of the variables. For example, dividing equation (3.23) by dy we obtain

$$\frac{\mathrm{d}z}{\mathrm{d}y} = \left(\frac{\partial z}{\partial x}\right)_y\frac{\mathrm{d}x}{\mathrm{d}y} + \left(\frac{\partial z}{\partial y}\right)_x. \tag{3.25}$$

The variables most often used in physical applications are p, v and T, two of which are independent due to the existence of the relating equation of state. If we choose to introduce a new variable, u say, into our relation $z = z(x, y)$, the new variable must be related to the existing ones to be useful. Dividing equation (3.23) by dy at constant u we obtain

$$\left(\frac{\partial z}{\partial y}\right)_u = \left(\frac{\partial z}{\partial x}\right)_y\left(\frac{\partial x}{\partial y}\right)_u + \left(\frac{\partial z}{\partial y}\right)_x. \tag{3.26}$$

This equation relates partial derivatives when the variable being held constant is changed.

Example 3.6
The molar heat capacity at constant pressure $C_p = (\partial H/\partial T)_p$, and that at constant volume $C_v = (\partial U/\partial T)_v$ are related because $H = U + pv$, so that

$$C_v = \left(\frac{\partial U}{\partial T}\right)_v = \left(\frac{\partial}{\partial T}(H - pv)\right)_v$$

$$= \left(\frac{\partial H}{\partial T}\right)_v - v\left(\frac{\partial p}{\partial T}\right)_v.$$

To develop this into a relation between C_p and C_v we need to relate $(\partial H/\partial T)_v$ and $(\partial H/\partial T)_p$. From equation (3.26)

$$\left(\frac{\partial H}{\partial T}\right)_v = \left(\frac{\partial H}{\partial T}\right)_p + \left(\frac{\partial H}{\partial p}\right)_T\left(\frac{\partial p}{\partial T}\right)_v$$

so that

$$C_v = C_p - \left(\frac{\partial p}{\partial T}\right)_v\left[v - \left(\frac{\partial H}{\partial p}\right)_T\right].$$

The term $(\partial H/\partial p)_T$ can be interpreted using the Gibbs equation

$$dH = T\,dS + v\,dp$$

and dividing by dp at constant T to obtain

$$\left(\frac{\partial H}{\partial p}\right)_T = T\left(\frac{\partial S}{\partial p}\right)_T + v.$$

Using the Maxwell relation of Example 3.4, we obtain

$$\left(\frac{\partial H}{\partial p}\right)_T = v - T\left(\frac{\partial v}{\partial T}\right)_p,$$

so that

$$C_p = C_v + T\left(\frac{\partial p}{\partial T}\right)_v\left(\frac{\partial v}{\partial T}\right)_p.$$

This is as far as thermodynamic and mathematical relations can take us. For a perfect gas, $pv = RT$ so that $(\partial v/\partial T)_p = R/p$ and $(\partial p/\partial T)_v = R/v$. Hence $(\partial H/\partial p)_T = 0$ and

$$C_p = C_v + R, \qquad \text{perfect gas.}$$

3.7 Extensive and intensive variables; Euler's theorem

The terms extensive and intensive are used in a physical context to distinguish between variables that are proportional to amount of substance (extensive) and those independent of amount of substance (intensive). Extensive variables include mass, volume, energy and entropy, and intensive variables include pressure, temperature and refractive index.

In mathematical terms this is a distinction between functions of first degree, and those of zero degree, in the amount of substance. A function $f(x)$ is of degree m in variable x if when x is multiplied by a constant, say λ, the value of the function is multiplied by λ^m,

$$f(\lambda x) = \lambda^m f(x).$$

A function of first degree is one that is proportional to the variable

$$f(\lambda x) = \lambda f(x) \qquad \text{for } m = 1, \text{ extensive.}$$

An intensive property, on the other hand, is a zero-degree function

$$f(\lambda x) = f(x) \qquad \text{for } m = 0, \text{ intensive.}$$

In physical problems we usually have more than one variable to consider, some of which will be extensive and some intensive. If an extensive variable $z(a, b, x, y)$ depends on extensive variables a, b and on intensive variables x, y, multiplying the amount of substance by λ means multiplying the extensive variables by λ but keeping the intensive variables constant. Then

$$z(\lambda a, \lambda b, x, y) = \lambda z(a, b, x, y).$$

Partial differentiation with respect to λ at constant a, b, x and y gives

$$\frac{\partial z}{\partial(\lambda a)}\frac{\mathrm{d}(\lambda a)}{\mathrm{d}\lambda} + \frac{\partial z}{\partial(\lambda b)}\frac{\mathrm{d}(\lambda b)}{\mathrm{d}\lambda} = z(a, b, x, y)$$

$$a\frac{\partial z}{\partial(\lambda a)} + b\frac{\partial z}{\partial(\lambda b)} = z(a, b, x, y).$$

This is true for any λ, so put $\lambda = 1$ to give

$$a\frac{\partial z}{\partial a} + b\frac{\partial z}{\partial b} = z. \qquad (3.27)$$

In general, a function $f(x_1, x_2, x_3, \dots)$ is homogeneous of degree m if

$$f(\lambda x_1, \lambda x_2, \lambda x_3, \dots) = \lambda^m f(x_1, x_2, x_3, \dots) \qquad (3.28)$$

and Euler's theorem is

$$x_1\frac{\partial f}{\partial x_1} + x_2\frac{\partial f}{\partial x_2} + x_3\frac{\partial f}{\partial x_3} + \dots = mf. \qquad (3.29)$$

In the above example the function is first-degree in the extensive variables but zero-degree in the intensive ones, so that only the former appear in equation (3.27)

When we have an equation in differential form for a function of state in terms of both extensive and intensive variables we may use Euler's theorem in the form (3.27).

Example 3.7

For a mixture of substances 1 and 2 which is able to gain or lose numbers of moles $\mathrm{d}n_1$ and $\mathrm{d}n_2$ of substance, the total energy U is given in differential form by

$$dU = T\,dS - p\,dv + \mu_1\,dn_1 + \mu_2\,dn_2.$$

The variables S, v, n_1 and n_2 are extensive. We write

$$U = U(S, v, n_1, n_2)$$

and Euler's theorem gives

$$S\frac{\partial U}{\partial S} + v\frac{\partial U}{\partial v} + n_1\frac{\partial U}{\partial n_1} + n_2\frac{\partial U}{\partial n_2} = U.$$

But from the given equation

$$\frac{\partial U}{\partial S} = T, \qquad \frac{\partial U}{\partial v} = -p, \qquad \frac{\partial U}{\partial n_1} = \mu_1 \quad \text{and} \quad \frac{\partial U}{\partial n_2} = \mu_2,$$

so that

$$U = TS - pv + n_1\mu_1 + n_2\mu_2.$$

On the other hand, for the Gibbs function G we have

$$dG = v\,dp - S\,dT + \mu_1\,dn_1 + \mu_2\,dn_2$$

where n_1 and n_2 are extensive but p and T are intensive. Euler's theorem therefore gives

$$G = n_1\mu_1 + n_2\mu_2,$$

terms in p and T not appearing.

By using Euler's theorem in this way we are, in effect, integrating differential equations.

3.8 Taylor's theorem in partial derivatives

Taylor's theorem is derived for functions of a single independent variable in Section 2.9.2. The treatment for more than one independent variable is analogous but with total derivatives being replaced by partial derivatives with respect to each independent variable. Thus if a function $f(x, y)$ is known at point (a, b) we assume that for the point $(a + x, b + y)$ the increment in the function may be written as a power series in the form

$$f(a + x, b + y) - f(a, b) = a_1 x + a_2 y + a_3 x^2 + a_4 xy + a_5 y^2 + \dots.$$

We denote the partial derivative with respect to x, $\partial f/\partial x$, by f_x and higher derivatives $\partial^2 f/\partial x^2$ by f_{xx}, $\partial^2 f/\partial x\,\partial y$ by f_{xy} and so on. Then

$$f_x(a + x, b + y) = a_1 + 2a_3 x + a_4 y + \dots$$

and putting $x = y = 0$ gives $f_x(a, b) = a_1$. Similarly $f_y(a, b) = a_2$. Taking second derivatives and then putting $x = y = 0$ gives $a_3 = f_{xx}(a, b)/2$ and $a_4 = f_{xy}(a, b)$. In this way we obtain Taylor's theorem in the form

$$f(a + x, b + y) = f(a, b) + xf_x + yf_y + \frac{1}{2!}(x^2 f_{xx} + 2xy f_{xy} + y^2 f_{yy}) + \ldots ,$$

$$(3.30)$$

where the derivatives are evaluated at the point (a, b). This is used in Section 7.5.

3.9 Vectors

A vector quantity is one that has magnitude, direction and sense, in contrast to scalar quantities, which have magnitude only. Mass and time are examples of scalar quantities, whilst displacement, velocity and force are vectors. A vector quantity may be represented by an arrow of length proportional to the magnitude of the quantity and in the appropriate direction and sense; they are then added together by joining the arrows head to tail as in Fig. 3.3, where displacements OP and PQ are added to give OQ.

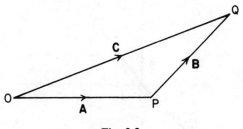

Fig. 3.3

We denote a vector quantity by a letter in bold-face type, such as **A**, and the (scalar) magnitude of **A** is denoted by A or $|\mathbf{A}|$. Then if OP, PQ and OQ represent the vectors **A**, **B** and **C** we can write the vector equations

$$\mathbf{A} + \mathbf{B} = \mathbf{C}, \qquad \mathbf{A} + \mathbf{B} - \mathbf{C} = 0, \qquad (3.31)$$

so that $-\mathbf{C}$ is a vector with the same magnitude and direction as **C** but opposite sense. When **C** is obtained by adding together **A** and **B**, it is called the resultant of the two vectors. Conversely, we may resolve vector **C** into its components, **A** and **B**, in the directions of **A** and **B**.

The magnitude s of vector **s** is a scalar, and may be operated upon like any other scalar quantity. Thus we may multiply vector **s** by scalar a to obtain a vector of magnitude as and in the same direction and sense as **s**. We may also differentiate with respect to a scalar quantity; differentiation of the scalar magnitude s of displacement **s** with respect to time t gives the velocity vector **v** in the same direction as **s** and of magnitude ds/dt. Arguments of this kind may be used to show that momentum, acceleration and force are also vector quantities.

A useful analytical device is to define a set of vectors **i**, **j** and **k** in the directions of rectangular coordinates in space and each of unit length. In Fig. 3.4 the line OP represents vector **A** and gives scalar projections A_x, A_y and A_z on the x, y- and z-axes. The product $\mathbf{i}A_x$ is then a vector of magnitude A_x in the direction of the x-axis, and this is the component of **A** in the x direction. We may then add the components to obtain

$$\mathbf{A} = \mathbf{i}A_x + \mathbf{j}A_y + \mathbf{k}A_z. \tag{3.32}$$

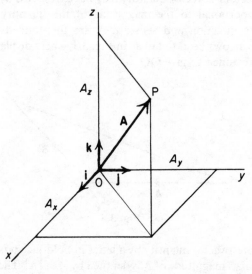

Fig. 3.4

Addition or subtraction of vectors is then achieved by adding or subtracting the components giving, for example,

$$\mathbf{A} - \mathbf{B} = \mathbf{i}(A_x - B_x) + \mathbf{j}(A_y - B_y) + \mathbf{k}(A_z - B_z). \tag{3.33}$$

When a body is free to rotate in space it does so about its centre of mass

(Section 2.11). We then define three rectangular coordinates through that centre of mass as origin, so that any arbitrary axis of rotation makes angles α, β and γ with the x-, y- and z-axes. It may be shown that a rotation through angle ω about that arbitrary axis is then the same as the sum of rotations ωl, ωm and ωn about the three coordinate axes, where l, m and n are the direction cosines (Section 1.7.3) of the axis of rotation. It follows from this that if we represent angular rotation by an arrow drawn in the axis of rotation, of length proportional to the angle of rotation and in the conventional sense that the rotation is in the direction of a right-hand screw, then the component rotations about the axes obey Pythagoras's equation

$$(\omega l)^2 + (\omega m)^2 + (\omega n)^2 = \omega^2 (l^2 + m^2 + n^2) = \omega^2.$$

and the arrows may be treated as any other vector quantity.

Using the same arguments about scalar manipulation as above, it follows that angular velocity, angular momentum (the moment of linear momentum) and torque (the moment of a force) may be represented by vectors in the same way. There is, however, one difference between the vectors used to represent these angular properties and the vector quantities previously considered, namely that the vector is localized in the axis of rotation. Vectors based on displacement of an extended body, on the other hand, are non-localized in the sense that they may be freely moved parallel to themselves.

Example 3.8
When a gyroscope spins with angular velocity ω about a horizontal axis it has angular momentum $I\omega$, represented by a vector located in the axis of rotation as shown in Fig. 3.5. If it is supported at one end only,

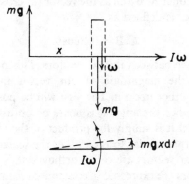

Fig. 3.5

the weight mg of the gyroscope is balanced by upthrust mg at the support; these forces generate a couple mgx represented by a vector in the horizontal plane perpendicular to the axis of rotation. From equation (2.70).

couple = rate of change of angular momentum,

so that in time interval dt angular momentum mgxdt is generated, represented by a vector into the plane of the paper in Fig. 3.5 in the direction perpendicular to the axis of rotation. The two angular momentum vectors are drawn in the figure as when viewed vertically downwards; the resultant, and so the axis of rotation of the gyroscope, rotates in the horizontal plane in an anticlockwise direction from that point of view. This rotation of the gyroscope about the vertical axis is called precession, the direction of precession being the opposite of what would occur if the gyroscope were a wheel running on a horizontal table.

This also applies to the effect of a magnetic field on a spinning magnetic dipole; the field imposes a couple on the spinning dipole, which therefore precesses about the direction of the magnetic field. Nuclei that behave in this way are detected by the absorption of radiation of frequency equal to the precession frequency in a nuclear magnetic resonance spectrometer, the precession frequency depending on the magnitude of the magnetic field strength at the particular nucleus.

3.9.1 Scalar and vector products

The product of two vectors, **A** and **B**, is defined in two ways, as the scalar (or dot) product **A** . **B** and as the vector (or cross) product **A** × **B**. The scalar product is defined as

$$\mathbf{A} . \mathbf{B} = AB\cos\theta, \qquad (3.34)$$

where θ is the angle between the two vectors. This product is a scalar quantity and is the magnitude of one vector multiplied by the projection of the other upon it. The sign will be positive or negative according to whether the angle θ is acute or obtuse. When the two vectors are parallel it is simply the product of their magnitudes, and when they are perpendicular to each other the scalar product is zero; such perpendicular vectors are called orthogonal.

When the vectors are expressed in component form, the unit vectors **i**, **j** and **k** are defined to be orthogonal so that **i**.**i** = **j**.**j** = **k**.**k** = 1 and

$\mathbf{i}.\mathbf{j} = \mathbf{j}.\mathbf{k} = \mathbf{k}.\mathbf{i} = 0$ and the scalar product becomes

$$\mathbf{A}.\mathbf{B} = (\mathbf{i}A_x + \mathbf{j}A_y + \mathbf{k}A_z).(\mathbf{i}B_x + \mathbf{j}B_y + \mathbf{k}B_z)$$
$$= A_x B_x + A_y B_y + A_z B_z. \tag{3.35}$$

The vector product or cross-product $\mathbf{A} \times \mathbf{B}$ is defined as a vector of magnitude $|\mathbf{A} \times \mathbf{B}|$ given by

$$|\mathbf{A} \times \mathbf{B}| = AB\sin\theta \tag{3.36}$$

in the direction perpendicular to the plane containing $\mathbf{A}$ and $\mathbf{B}$ and in the sense that a right-hand screw brings one from $\mathbf{A}$ to $\mathbf{B}$ through an angle of less than π. When the order of writing the product is reversed, the direction of the vector reverses so that $\mathbf{B} \times \mathbf{A} = -(\mathbf{A} \times \mathbf{B})$.

The vector product represents the magnitude of one vector multiplied by the component of the other perpendicular to the first, which is the area of the parallelogram based on the two vectors as in Fig. 3.6.

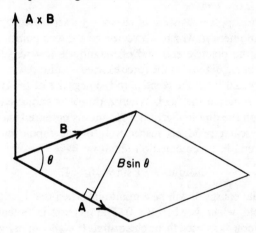

Fig. 3.6

This illustrates another useful principle, that plane surface area may be represented by a vector of length proportional to the area and in the direction normal to the surface. When plane surface area A is viewed obliquely from the direction making an angle θ to the normal to the surface, the area seen is $A \cos \theta$. If we define the line of sight by unit vector $\mathbf{r}$ this may be expressed in the form of a scalar product:

$$\text{effective area} = \mathbf{A}.\mathbf{r} = A \cos \theta. \tag{3.37}$$

The moment of a vector about a point, such as that of force $\mathbf{F}$ about O in Fig. 3.7, can be expressed as the vector product of $\mathbf{F}$ and the radius

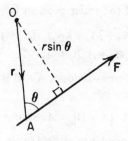

Fig. 3.7

vector $\mathbf{r}$ of any point A on the line of $\mathbf{F}$, since

$$\text{moment of force} = \mathbf{F} \times \mathbf{r} = Fr \sin \theta \qquad (3.38)$$

in the direction perpendicular to both $\mathbf{F}$ and $\mathbf{r}$ and in the sense of the right-hand screw convention.

An electric dipole consisting of charges $\pm e$ separated by distance r has dipole moment $\boldsymbol{\mu}$, which is a vector on the axis pointing from the negative to the positive end and of magnitude $\mu = re$. Electric field strength $\mathbf{E}$, being defined as the force exerted on unit positive charge, is a vector directed from the positive to the negative of the two charged plates that generate the field. When a dipole is subjected to a field, the torque on the dipole (Section 2.11) and its potential energy are the vector and scalar products respectively of the dipole moment and the field strength. From equation (2.68) we have

$$\text{torque} = \mu E \sin \theta = \boldsymbol{\mu} \times \mathbf{E}. \qquad (3.39)$$

The potential energy V will be a minimum when the dipole is aligned with the field, which is when $\theta = 0$. The increase in potential energy when the dipole is rotated through an angle θ as shown in Fig. 3.8 is the work done against the forces $\pm Ee$ acting at the ends of the dipole. Each end moves a distance $x = (r - r \cos \theta)/2$ against the direction of the field so that, anticipating our discussion of integration,

$$\mathrm{d}x = \tfrac{1}{2} r \sin \theta \, \mathrm{d}\theta,$$

$$V = 2 \int_0^x Ee \, \mathrm{d}x = Eer \int_0^\theta \sin \theta \, \mathrm{d}\theta = -\mu E \cos \theta,$$

so that

$$V = -\boldsymbol{\mu} \cdot \mathbf{E} \qquad (3.40)$$

The condition that a scalar product of zero means that the vectors are

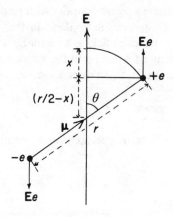

Fig. 3.8

orthogonal (perpendicular to each other) may be used to explain the general definition of orthogonal functions. We may extend equation (3.35) by writing

$$\mathbf{A} . \mathbf{B} = \sum A_i B_i, \tag{3.41}$$

where A_i is the component of $\mathbf{A}$ in the ith coordinate direction. If we then allow the number of dimensions to tend to infinity we are supposing $\mathbf{A}$ to be made up of an infinite number of components, or $\mathbf{A}$ may be replaced by a continuous function $f(x)$. Under these conditions we may replace the summation in (3.41) by integration over the continuous variable, to give

$$\int_a^b f(x)g(x)\mathrm{d}x = 0 \tag{3.42}$$

as the condition for the functions $f(x)$ and $g(x)$ to be orthogonal in the interval (a, b). This is used in wave mechanics when we replace $f(x)$ by the wavefunction ψ, and $g(x)$ by the complex conjugate ψ^* (Section 1.10) so as to obtain the condition for ψ and ψ^* to be orthogonal.

3.9.2 Scalar and vector fields

A region of three-dimensional space in which a quantity varies from point to point is called a field. If the quantity is scalar, such as density, temperature, pressure or concentration, we have a scalar field and the normal methods of calculus apply. A gravitational, electric or magnetic field, on the other hand, is a vector field because it has direction and sense, as well as magnitude, at any point.

When we differentiate a scalar quantity with respect to a coordinate direction in space we obtain the gradient in that direction. We can express the gradient of a scalar field as a vector by writing, for example, $i\partial\phi/\partial x$ for the gradient of $\phi(x, y, z)$ in the x direction, where i is a unit vector in that direction. In three dimensions we may write

$$\mathbf{grad}\ \phi = \mathbf{i}\frac{\partial\phi}{\partial x} + \mathbf{j}\frac{\partial\phi}{\partial y} + \mathbf{k}\frac{\partial\phi}{\partial z}. \tag{3.43}$$

This may also be written in operator form; we define 'nabla' as the operator

$$\nabla = \mathbf{i}\frac{\partial}{\partial x} + \mathbf{j}\frac{\partial}{\partial y} + \mathbf{k}\frac{\partial}{\partial z}, \tag{3.44}$$

so that (3.43) becomes

$$\nabla\phi = \mathbf{grad}\,\phi. \tag{3.45}$$

The significance of this is that by adding together the gradients in the three coordinate directions as vectors we obtain the magnitude and direction in which the rate of change in ϕ is greatest; $\mathbf{grad}\ \phi$ is a vector that is normal to the contour surfaces over which ϕ is constant. Thus if ϕ is the (scalar) potential of mass m in the earth's gravitational field, and z is the vertical axis, we have $\partial\phi/\partial x = \partial\phi/\partial y = 0$ and $\partial\phi/\partial z = mg$, so that $\mathbf{grad}\ \phi = kmg$, meaning that the maximum, in this case the only, potential gradient is mg in the vertical direction. This notation becomes useful in more complicated cases; the direction and strength of the electric field at a point due to a surrounding distribution of charges in space can be found as $-\mathbf{grad}\ \psi$ where ψ is the total potential due to the superimposed charges. Gradients of temperature, concentration or electric potential will produce a flow of heat, diffusion of material or an electric current, so that this notation is useful in three-dimensional problems of these kinds.

When we are concerned with vector quantities in three-dimensional space, new possibilities arise. In a fluid (liquid or gas) a flow of material can occur either in parallel streamlines or in the form of eddies or swirls. If we consider the work done in moving from one point to another in space, when this is independent of the path taken between the points we say that we have a conservative field. This applies to mass in a gravitational field if we neglect air resistance.

If the field exerts force $\mathbf{F}$ on a particle, the work done in a displacement $d\mathbf{r}$ is the scalar product $\mathbf{F}.d\mathbf{r}$ given by

$$\text{work} = \mathbf{F}.d\mathbf{r} = F_x dx + F_y dy + F_z dz. \tag{3.46}$$

This work will be independent of path and the field will be conservative if (3.46) is an exact differential and the cross-differentiation identity applies (Section 3.5), or if

$$\frac{\partial F_x}{\partial y} = \frac{\partial F_y}{\partial x}, \qquad \frac{\partial F_x}{\partial z} = \frac{\partial F_z}{\partial x} \qquad \text{and} \qquad \frac{\partial F_z}{\partial y} = \frac{\partial F_y}{\partial z}. \qquad (3.47)$$

We may therefore express departure from the condition of the work being independent of path as departure from (3.47). This is called the curl of the vector field, defined by

$$\text{curl } \mathbf{F} = \mathbf{i}\left(\frac{\partial F_z}{\partial y} - \frac{\partial F_y}{\partial z}\right) + \mathbf{j}\left(\frac{\partial F_x}{\partial z} - \frac{\partial F_z}{\partial x}\right) + \mathbf{k}\left(\frac{\partial F_y}{\partial x} - \frac{\partial F_x}{\partial y}\right).$$
$$(3.48)$$

The first bracket in (3.48) is the x component of the vector **curl F**, and it depends on the changes in **F** in the (y, z) plane, and in fact measures rotation of the field in that plane. The significance of **curl F** is that it measures eddies or swirls in the field: if we move from one point to another along a path that is opposed to the eddies we do more work than if we move along a path that is in the same direction as the eddies.

Equation (3.48) contains the derivatives of the components of **F** in perpendicular directions, such as $\partial F_z/\partial x$. The derivatives in the parallel directions, such as $\partial F_x/\partial x$, give the components of the increase in magnitude of the vector. This is a scalar quantity called divergence, defined by

$$\text{div } \mathbf{F} = \frac{\partial F_x}{\partial x} + \frac{\partial F_y}{\partial y} + \frac{\partial F_z}{\partial z}, \qquad (3.49)$$

and detects sources or sinks in the field; this would apply if heat were being supplied at a point in a flowing fluid because thermal expansion would then create increase in velocity and so a positive divergence of the velocity field.

These concepts provide concise notation in problems involving three-dimensional fields. The nabla operator (3.44) can be used for vector fields by noting that it can be regarded as a vector, so that both scalar and vector products of ∇ with **F** can be written. It can be seen from the above definitions that

$$\nabla . \mathbf{F} = \text{div } \mathbf{F} \qquad \text{and} \qquad \nabla \times \mathbf{F} = \text{curl } \mathbf{F}. \qquad (3.50)$$

CHAPTER 4

Integration

4.1 Significance and notation

Integration can be defined as the inverse of differentiation or as the limit of a sum. If the differential of a function $f(x)$ is $f'(x)\,dx$, then the integral of $f'(x)\,dx$ is $f(x)$:

$$\int f'(x)\,dx = f(x) + C. \tag{4.1}$$

This is called an indefinite integral because the addition of any constant c to the integral $f(x)$ will give the same differential $f'(x)\,dx$. Also, if we multiply $f(x)$ by a constant, K, then $f'(x)\,dx$ is multiplied by the same constant; this means that constants can be taken outside the integral sign:

$$\int Kf'(x)\,dx = K\int f'(x)\,dx + C. \tag{4.2}$$

When the integration is between particular values of x, say from $x = a$ to $x = b$, we have a definite integral and the integration constant cancels out.

$$\int_a^b f'(x)\,dx = \left[\,f(x)\,\right]_a^b = f(b) - f(a). \tag{4.3}$$

This definition of integration as the inverse of differentiation gives the rules for integrating simple functions, and these are discussed in the following sections.

A definite integral can also be defined as the limit of a sum, which is the reason for the use of the integral sign:

$$\int_a^b f(x)\,\mathrm{d}x = \lim_{\delta x \to 0} \sum_{x=a}^{x=b} f(x)\,\delta x, \tag{4.4}$$

where Σ denotes the sum produced by adding successive increments δx to x, starting at $x = a$ and ending at $x = b$.

The proof of the equivalence of these two definitions requires careful mathematical analysis, but a geometrical interpretation is shown in Fig. 4.1. The area of the shaded rectangle is $y\delta x$ and the sum of the areas of such rectangles becomes the area under the curve as $\delta x \to 0$.

$$\text{area } ab\text{BA} = \int_a^b y\,\mathrm{d}x = \int_a^b f(x)\,\mathrm{d}x. \tag{4.5}$$

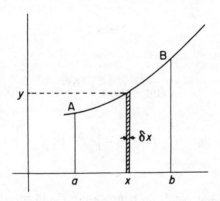

Fig. 4.1

Another interpretation is illustrated in Fig. 4.2. We can follow the curve $y = f(x)$ starting from the point A, which is $(a, f(a))$, by making small steps δx in x. A tangent to the curve has slope $\mathrm{d}y/\mathrm{d}x$, so that the increment δy in y along the tangent for increment δx in x is

$$\delta y = \frac{\mathrm{d}y}{\mathrm{d}x}\,\delta x. \tag{4.6}$$

In the limit as $\delta x \to 0$, a succession of such steps along tangents will follow the curve, and the sum of the steps between a and b becomes

$$\lim_{\delta x \to 0} \sum_{x=a}^{x=b} \frac{\mathrm{d}y}{\mathrm{d}x}\,\delta x = \int_a^b \frac{\mathrm{d}y}{\mathrm{d}x}\,\mathrm{d}x = \int_{f(a)}^{f(b)} \mathrm{d}y = f(b) - f(a). \tag{4.7}$$

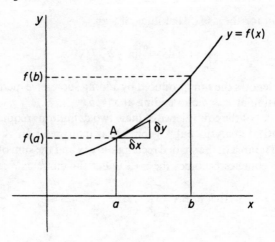

Fig. 4.2

Thus

$$f(b) = f(a) + \int_a^b \frac{dy}{dx}\, dx \tag{4.8}$$

or

$$f(b) = f(a) + \int_a^b f'(x)\, dx. \tag{4.9}$$

Example 4.1

If a quantity X depends on temperature T and pressure p, we write $X = X(T, p)$. The change in X from temperature T_1 to temperature T_2 at constant pressure is

$$X(T_2, p) - X(T_1, p) = \int_{T_1}^{T_2} \left(\frac{\partial X}{\partial T}\right)_p dT.$$

If X is the enthalpy H then $(\partial H/\partial T)_p$ is the heat capacity C_p and

$$H(T_2, p) - H(T_1, p) = \int_{T_1}^{T_2} C_p\, dT.$$

These interpretations of integration may be summarized in the form that: if we begin by knowing the properties at a point (the values of x_0 and dy/dx), integration produces a curve; if we know a curve ($y = f(x)$), integration produces an area; and if we know an area, integration will produce a volume.

It is assumed here that the function to be integrated is well-behaved in the sense that it is continuous and single-valued. Continuity usually applies in physical applications, but functions that are not single-valued are often met; these are discussed in Section 5.1.

4.2 Standard methods of integration

4.2.1 Simple functions

The integral of a function is that quantity which, when differentiated, produces the original function. Thus

$$\int x^n \, dx = \frac{x^{n+1}}{n+1} + C, \qquad n \neq -1,$$

$$\int dx = x + C,$$

$$\int e^x \, dx = e^x + C,$$

$$\int e^{ax} \, dx = \frac{1}{a} e^{ax} + C,$$

$$\int [f(x) + g(x)] \, dx = \int f(x) \, dx + \int g(x) \, dx.$$

4.2.2 Reciprocal functions

When $n = -1$ in the first example above, the rule does not apply. This particular case is evaluated by the following device:

$$\text{put} \quad y = \ln x, \qquad \text{then} \quad x = e^y,$$

$$\frac{dx}{dy} = e^y = x.$$

Taking reciprocals

$$\frac{dy}{dx} = \frac{1}{x}.$$

Hence

$$\int \frac{1}{x} \, dx = \int \frac{dy}{dx} \, dx = \int dy = y + C = \ln x + C.$$

4.2.3   *Integration by parts*

This applies to the integral of a product of two functions. From the differential of a product we have

$$d(uv) = u\,dv + v\,du.$$

We can integrate term by term to give

$$\int d(uv) = \int u\,dv + \int v\,du.$$

But

$$\int d(uv) = uv,$$

so that

$$\int u\,dv = uv - \int v\,du.$$

If we denote a function $f(x)$ by u and a derivative $g(x)\,dx$ by dv this gives

$$\int f(x)g(x)\,dx = f(x)\int g(x)\,dx - \int f'(x)\left(\int g(x)\,dx\right)dx.$$

It is useful also to express this in words, as 'the integral of a product of two functions is the first function times the integral of the second, minus the integral of (derivative of first function times integral of second)'.

Example 4.2

$$\int xe^{ax}\,dx = x\frac{1}{a}e^{ax} - \int \frac{1}{a}e^{ax}\,dx$$

$$= \frac{1}{a}e^{ax}\left(x - \frac{1}{a}\right) + C.$$

This can be used to illustrate an important principle, that the factors of the integrand must be chosen so that differentiation of one part and integration of the other shall yield a simple integral. This may not be the case if the parts are incorrectly chosen; we might wrongly regard the above example as a product of terms taken in the opposite order, as

$$\int e^{ax} x\,dx = e^{ax}\frac{x^2}{2} - \int ae^{ax}\frac{x^2}{2}\,dx$$

$$= \frac{x^2}{2}e^{ax} - \frac{a}{2}\int x^2\,e^{ax}\,dx,$$

thus producing only a more complicated integral. If this happens, the rule is to try again with the factors the other way round.

Example 4.3

Integration by parts can sometimes yield the same integral again, which can then be evaluated by the solution of an algebraic equation. Denoting the integral by the symbol I, we could have

$$I = \int \sin^2 \theta \, d\theta = -\sin \theta \cos \theta + \int \cos^2 \theta \, d\theta$$

$$= -\sin \theta \cos \theta + \int (1 - \sin^2 \theta) \, d\theta$$

$$= -\sin \theta \cos \theta + \int d\theta - I.$$

Hence

$$2I = \theta - \sin \theta \cos \theta + c,$$

$$I = \frac{\theta}{2} - \frac{1}{2} \sin \theta \cos \theta + c$$

so that, for example

$$\int_0^{\pi/2} \sin^2 \theta \, d\theta = \frac{\pi}{4}.$$

When a definite integral is integrated by parts the application of the limits is straightforward so long as we remember that the limits mean taking the value of the integral at the upper limit minus that at the lower limit. Thus

$$\int_\alpha^\beta u \, dv = \left[uv \right]_\alpha^\beta - \int_\alpha^\beta v \, du.$$

Example 4.4

The integral of ln x is obtained as the product of ln x and unity:

$$\int \ln x \, dx = \int (\ln x).1 \, dx$$

$$= x \ln x - \int \frac{1}{x} x \, dx$$

$$= x \ln x - \int dx$$

$$= x \ln x - x + C.$$

4.2.4　Integration by substitution

This applies when the integral can be regarded as the product or ratio of two parts, one part being the differential of the other part.

Example 4.5

$$I = \int \frac{x \, dx}{x^2 + 4}$$

Since $d(x^2 + 4) = 2x \, dx$, the factor of 2 being immaterial, we substitute

$$u = x^2 + 4, \qquad du = 2x \, dx.$$

Then

$$I = \frac{1}{2} \int \frac{du}{u} = \tfrac{1}{2} \ln u + C$$

$$= \tfrac{1}{2} \ln(x^2 + 4) + C.$$

Example 4.6

$$I = \int \frac{\ln x}{x} \, dx.$$

This is the product of $1/x$ and $\ln x$, where $d \ln x = (1/x) \, dx$. We therefore substitute $u = \ln x$, $du = dx/x$,

$$I = \int u \, du = \tfrac{1}{2} u^2 = \tfrac{1}{2} (\ln x)^2 + C.$$

When substitution is used to evaluate a definite integral, the change in variable means that the limits of integration must also be changed to the values of the new variable at those limits. Thus if Example 4.5 were between limits of 0 and 2 for x, the limits for u become the values of $u = x^2 + 4$ at $x = 0$ and $x = 2$, or $u = 4$ and $u = 8$. This can, of course, be avoided by returning to the original variable before taking the limits.

4.2.5　Expansion of algebraic functions

When

$$I = \int \frac{f(x)}{g(x)} \, dx,$$

where $f(x)$ is of higher degree (contains higher powers of x) than $g(x)$, the expression can be expanded by division of the denominator into the numerator.

Example 4.7

$$I = \int \frac{x^3}{x^2+2} \, dx = \int \left(x - \frac{2x}{x^2+2} \right) dx$$

$$= \int x \, dx - \int \frac{2x}{x^2+2} \, dx$$

$$= \frac{x^2}{2} - \ln(x^2+2) + C.$$

4.2.6 Integration by partial fractions

Algebraic functions with denominators that can be factorized can be expanded by resolution into partial fractions. This is the inverse of the simple addition of fractions; since

$$\frac{1}{x+2} + \frac{1}{x-1} = \frac{2x+1}{(x+2)(x-1)}$$

we can equally well resolve the right-hand side of this equality into the left. We define parameters A and B by writing

$$\frac{2x+1}{(x+2)(x-1)} = \frac{A}{x+2} + \frac{B}{x-1} = \frac{A(x+1)+B(x+2)}{(x+2)(x-1)},$$

which will be true if

$$A(x+1) + B(x+2) = 2x+1.$$

Since this must be true for any value of x, we can find A and B either by suitable choice of x to eliminate one of the parameters, or by equating constant terms and the coefficients of x. Thus

$$\text{if} \quad x = 1, \qquad 3B = 2, \qquad \text{and so} \quad B = 1,$$
$$\text{if} \quad x = -2, \quad -3A = -3, \qquad \text{and so} \quad A = 1.$$

The rules for obtaining partial fractions depend on the factors in the denominator of the expression;

(a) for factor $(x+a)$

$$\text{use} \quad \frac{A}{x+a}$$

(b) for factor $(x+a)^2$

$$\text{use} \quad \frac{A}{x+a} + \frac{B}{(x+a)^2}$$

(c) for factor $(ax^2 + bx + c)$

$$\text{use} \quad \frac{Ax + B}{ax^2 + bx + c}$$

Example 4.8

$$I = \int \frac{8 \, dx}{x^3 - 4x}$$

Put

$$\frac{8}{x^3 - 4x} = \frac{8}{x(x+2)(x-2)} = \frac{A}{x} + \frac{B}{x+2} + \frac{C}{x-2}$$

Then

$$A(x+2)(x-2) + Bx(x-2) + Cx(x+2) = 8.$$

Equating coefficients of x^2, of x and constant terms gives

$$A + B + C = 0, \quad B = C \quad \text{and} \quad A = -2.$$

Hence

$$B = C = 1,$$

and so

$$I = -2 \int \frac{dx}{x} + \int \frac{dx}{x+2} + \int \frac{dx}{x-2}$$

$$= \ln\left(\frac{x^2 - 4}{x^2}\right) + C.$$

4.3   Standard forms of integral and numerical methods.

Although it is generally possible to differentiate a continuous function, integration only gives a closed expression in particular cases.

Extensive tables are available giving the integral of many standard functions; a particularly comprehensive set is included in the *Handbook of Chemistry and Physics*, published at intervals by the Chemical Rubber Publishing Company, USA. One important standard form is

$$\int_0^\infty e^{-\alpha^2 x^2} \, dx = \frac{\sqrt{\pi}}{2\alpha}$$

and this is discussed in Section 5.2.

For integrals which cannot be evaluated in closed analytical form, geometrical and numerical methods are available at least to obtain an approximate solution. An integral can be interpreted as the area under a curve, so that we can draw the curve and measure the area between particular limits to evaluate the definite integral. This applies also when we want to integrate a quantity which appears as a curve on a recorder trace. The area can be found by counting squares or vertical strips, or by cutting out the shape and weighing it. Numerical methods are based on the same principle, and details are given in texts on numerical analysis. The Euler–Maclaurin formula is particularly useful and this is discussed in Section 4.6.

4.4 Multiple integration

Just as we may repeat the operation of differentiation to obtain higher-order derivatives, so we may repeat integration to evaluate multiple integrals. The notation used is to show the required number of integral signs together with the variables of integration. Thus

$$\int \int f(x) \, dx \, dx \qquad \text{or} \qquad \int dx \int f(x) \, dx,$$

both of which mean integrate with respect to x and then integrate the result again with respect to x. This is a double integral and the solution will contain two integration constants.

Multiple integration is most useful when we have more than one independent variable. If we write

$$\int f(x, y) \, dx$$

with two variables x and y in the integrand, it cannot be evaluated without a known relation between x and y so as to be able to substitute for y to obtain an integrand containing x only. An alternative, analogous to partial differentiation, is to integrate with respect to one variable whilst regarding other variables as constants. We do this when defining the multiple integral

$$\int_a^b \int_c^d f(x, y) \, dx \, dy \qquad \text{or} \qquad \int_a^b dy \int_c^d f(x, y) \, dx,$$

both of which mean integrate first with respect to x, keeping y constant, and then with respect to y, keeping x constant. The limits c and d apply to x and the limits a and b apply to y. Multiple integrals are evaluated from the inside to the outside in the first form shown, or from right to left in the second form. This order of integration does not, in fact, matter so long as the limits of integration for one variable do not contain the other variable.

Example 4.9

$$\int_1^3 \int_0^2 x^2 y \, dx \, dy = \int_1^3 \left[\frac{x^3 y}{3} \right]_0^2 dy = \int_1^3 \frac{8}{3} y \, dy$$

$$= \frac{8}{3} \left[\frac{y^2}{2} \right]_1^3 = \frac{32}{3}.$$

Notice that the limits of integration must be applied successively, not simultaneously.

Example 4.10

$$\int_0^1 \int_0^y x^2 y \, dx \, dy = \int_0^1 \left[\frac{x^3 y}{3} \right]_0^y dy = \int_0^1 \frac{y^4}{3} \, dy$$

$$= \left[\frac{y^5}{15} \right]_0^1 = \frac{1}{15}.$$

Since the variable y occurs in the limits of the integration with respect to x it is essential that the inner integral be evaluated first. If the integral were written in reverse order we would have

$$\int_0^y \int_0^1 x^2 y \, dy \, dx = \int_0^y \left[\frac{x^2 y^2}{2} \right]_0^1 dx = \int_0^y \frac{x^2}{2} \, dx$$

$$= \left[\frac{x^3}{6} \right]_0^y = \frac{y^3}{6}.$$

4.5   Differentiation of integrals; Leibnitz's theorem

Differentiation is the inverse of integration so that the operations of differentiation and integration, both with respect to the same variable, will cancel.

$$\frac{d}{dx} \int f(x) \, dx = \int \frac{df(x)}{dx} \, dx = \int df(x) = f(x).$$

If, however, we have a function of more than one independent variable we may differentiate with respect to one variable and integrate with respect to another. Since the variables are independent, these operations are independent of each other, so that the order of differentiation and integration may be reversed.

$$\frac{\partial}{\partial y} \int f(x, y)\, dx = \int \frac{\partial f(x, y)}{\partial y}\, dx. \qquad (4.10)$$

This will have meaning when the integral is between limits, and complications arise if the limits of integration contain the variable of differentiation in which case Leibnitz's theorem is used, which is

$$\frac{\partial}{\partial y} \int_{a(y)}^{b(y)} f(x, y)\, dx = \int_{a(y)}^{b(y)} \frac{\partial f(x, y)}{\partial y}\, dx + f(b, y)\frac{\partial b}{\partial y} - f(a, y)\frac{\partial a}{\partial y}. \qquad (4.11)$$

Example 4.11
Leibnitz's theorem can be used to obtain an alternative form for an integral by differentiation with respect to a parameter other than the variable of integration. Thus if we differentiate the integral

$$\int_0^\infty x e^{-ax^2}\, dx = \frac{1}{2a}$$

with respect to a we obtain

$$-\int_0^\infty x^3 e^{-ax^2}\, dx = \frac{d}{da}\frac{1}{2a} = -\frac{1}{2a^2}.$$

4.6 The Euler–Maclaurin theorem

When we are concerned with a function for which no simple closed form of integral exists, we may be able to expand the function as a series and then integrate term by term. This will apply so long as the function and the series used both converge sufficiently rapidly in the interval considered. The Euler–Maclaurin theorem can then be used, and is derived here by the use of Taylor's theorem.

We wish to obtain a series expansion of the definite integral

$$\int_{x_0}^{x_1} f(x)\, dx$$

so that the integral may then be evaluated. We first expand $f(x)$ as a

Taylor series about the value $f(\alpha)$ at $x = \alpha$ in the form

$$f(x) = f(\alpha) + (x - \alpha)f'(\alpha) + \frac{(x - \alpha)^2}{2!}f''(\alpha) + \ldots, \qquad (4.12)$$

and then integrate to obtain

$$\int_{x_0}^{x_1} f(x)\,dx = (x_1 - x_0)f(\alpha) + \frac{(x_1 - \alpha)^2 - (x_0 - \alpha)^2}{2!}f'(\alpha)$$

$$+ \frac{(x_1 - \alpha)^3 - (x_0 - \alpha)^3}{3!}f''(\alpha) + \ldots,$$

If we now choose α to be the mid point of the interval (x_0, x_1) the terms containing derivatives of odd order will cancel; we put $h = (x_1 - x_0)$, so that $(x_1 - \alpha) = h/2$ and $(x_0 - \alpha) = -h/2$, giving

$$\int_{x_0}^{x_1} f(x)\,dx = hf(\alpha) + \frac{h^3}{2^2 3!}f''(\alpha) + \frac{h^5}{2^4 5!}f^4(\alpha) + \ldots. \qquad (4.13)$$

We now use Taylor series again to express $f(\alpha), f''(\alpha), f^4(\alpha), \ldots$ in terms of the values at the end points of the interval; we adopt the device of taking the sum of the values of the function at the end points, but the differences of the values of the derivatives. Again using (4.12), we have

$$f(\alpha) = f(x) - (x - \alpha)f'(\alpha) - \frac{(x - \alpha)^2}{2!}f''(\alpha) - \ldots.$$

When we substitute $x = x_0$ and $x = x_1$ and add the resulting expressions, the terms containing odd powers of $(x - \alpha)$ cancel since $(x_1 - \alpha) = -(x_0 - \alpha)$, and we obtain

$$f(\alpha) = \frac{1}{2}\left(f(x_1) + f(x_0)\right) - \frac{h^2}{2^2 2!}f''(\alpha) - \frac{h^4}{2^4 4!}f^4(\alpha) - \ldots, \qquad (4.14)$$

and this is used to eliminate $f(\alpha)$ from (4.13). We also write Taylor expansions of $f'(x), f^3(x), \ldots$ in the form

$$f'(x) = f'(\alpha) + (x - \alpha)f''(\alpha) + \frac{(x - \alpha)^2}{2!}f^3(\alpha) + \ldots,$$

$$f^3(x) = f^3(\alpha) + (x - \alpha)f^4(\alpha) + \frac{(x - \alpha)^2}{2!}f^5(\alpha) + \ldots.$$

By taking the differences between the values of these expressions at the limits, the constant terms and those containing even powers of $(x - \alpha)$

will cancel to leave

$$f''(\alpha) = \frac{1}{h}\left(f'(x_1) - f'(x_0) \right) - \frac{h^2}{2^2 3!} f^4(\alpha) - \ldots , \qquad (4.15)$$

$$f^4(\alpha) = \frac{1}{h}\left(f^3(x_1) - f^3(x_0) \right) - \frac{h^2}{2^2 3!} f^6(\alpha) - \ldots , \qquad (4.16)$$

and so on. Substitution of (4.14), (4.15) and (4.16) into (4.13) gives

$$\int_{x_0}^{x_1} f(x)\,\mathrm{d}x = \frac{1}{2} h \left(f(x_1) + f(x_0) \right)$$

$$- \frac{1}{2^2}\left(\frac{1}{2!} - \frac{1}{3!} \right) h^2 \left(f'(x_1) - f'(x_0) \right)$$

$$+ \frac{1}{2^4}\left(\frac{1}{3!}\left(\frac{1}{2!} - \frac{1}{3!} \right) - \left(\frac{1}{4!} - \frac{1}{5!} \right) \right) h^4 \left(f^3(x_1) - f^3(x_0) \right)$$

$$- \text{terms in } \left(f^5(x_1) - f^5(x_0) \right), \text{ etc.} \qquad (4.17)$$

This formula can be used to evaluate the integral in terms of the values of the function $f(x)$ and of the derivatives of odd order $f'(x), f''(x), \ldots$, all at the end points of the integration range $x = x_0$ and $x = x_1$. The advantage of this form of equation is that we can divide the range of integration into small parts, which we show symbolically as

$$\int_{x_0}^{x_n} f(x)\,\mathrm{d}x = \int_{x_0}^{x_1} + \int_{x_1}^{x_2} + \ldots + \int_{x_{n-1}}^{x_n},$$

and the right hand side of (4.17) then simplifies. The first term can be expressed in terms of the sum of the values of $f(x)$ at the n points; in particular, if $h = 1$ this gives

$$\frac{1}{2} \sum_{i=0}^{n} \left(f(x_i) + f(x_{i+1}) \right) = \sum_{i=0}^{n} f(x_i) - \frac{1}{2}\left(f(x_0) + f(x_n) \right),$$

and in the terms containing the derivatives, the intermediate terms cancel to leave only the values at the ends of the whole range. Hence we obtain the Euler–Maclaurin formula in the form

$$\int_{x_0}^{x_n} f(x)\,\mathrm{d}x = \sum_{i=0}^{n} f(x_i) + \frac{1}{2}\left(f(x_n) - f(x_0) \right)$$

$$- \frac{1}{12}\left(f'(x_n) - f'(x_0) \right)$$

$$+ \frac{1}{720}\left(f^3(x_n) - f^3(x_0) \right) - \ldots , \qquad (4.18)$$

and the numerical coefficients of subsequent terms can be shown to be $1/30240$, $1/1209600$, ...

The two most common applications of this formula are to the numerical evaluation of convergent, but intractable, integrals which do not possess a solution in a closed form (numerical quadrature), and to the replacement of the sum of a convergent but intractable series by an integral (e.g. in statistical mechanics).

The formula as written in (4.21) contains the derivatives at both ends of the range of integration; if this range is from zero to infinity and if the values of the derivatives approach zero as x tends to infinity, the formula simplifies to

$$\int_0^\infty f(x)\,dx = \sum_{i=0}^\infty f(x_i) - \frac{1}{2}f(0) + \frac{1}{12}f'(0) - \frac{1}{720}f^3(0)$$

$$+ \frac{1}{30240}f^5(0) - \ldots \qquad (4.22)$$

Example 4.12
Use the Euler–Maclaurin theorem to evaluate $\int_0^\infty e^{-\alpha^2 x^2}\,dx$ when $\alpha = 0.5$, and compare the result with example 5.1.

We put $f(x) = e^{-\alpha^2 x^2}$, from which we obtain $f(0) = 1, f(\infty) = 0$, and all of the derivatives of odd order are zero at both limits. Hence by (4.22),

$$\int_0^\infty e^{-\alpha^2 x^2}\,dx = \sum_{n=0}^\infty e^{-\alpha^2 n^2} - \frac{1}{2}.$$

The integral is shown in example 5.1 to have the value $\sqrt{\pi}/2\alpha$. In order to obtain rapid convergence we must have a small enough value of α. Putting $\alpha = 0.5$ we obtain

$$\int_0^\infty e^{-\alpha^2 x^2}\,dx = \frac{\sqrt{\pi}}{2\alpha} = 1.772454,$$

and direct summation of the series gives

$$\sum_{n=0}^\infty e^{-\alpha^2 n^2} = 1 + e^{-0.5} + e^{-1} + e^{-2.25} + e^{-4} + \ldots.$$

The sum of this series to eight terms is 2.272454, so that we have agreement with the Euler–Maclaurin formula.

This expression arises in calculating the contribution from translation in a particular direction to the partition function in statistical mechanics. We then require to evaluate the sum

$$\sum_{n=1}^{\infty} e^{-n^2 h^2/8ma^2 kT} = \sum_{n=0}^{\infty} e^{-n^2 h^2/8ma^2 kT} - 1.$$

The Euler–Maclaurin formula gives

$$\sum_{n=0}^{\infty} e^{-n^2 h^2/8ma^2 kT} = \int_0^{\infty} e^{-n^2 h^2/8ma^2 kT} + \frac{1}{2}$$

$$= \left(\frac{2\pi mkT}{h^2}\right)^{1/2} a + \frac{1}{2},$$

so that

$$\sum_{n=1}^{\infty} e^{-n^2 h^2/8ma^2 kT} = \left(\frac{2\pi mkT}{h^2}\right)^{1/2} a - \frac{1}{2}.$$

Since the value of the integral is usually a very large number, the correction of $-1/2$ given by the Euler–Maclaurin formula may usually be neglected and we say that the sum may be replaced by the integral for the purposes of evaluation.

Example 4.13
Following the previous example, the relation which corresponds to the rotational contribution to the partition function is

$$\sum_{J=0}^{\infty} (2J+1)e^{-J(J+1)\theta/T},$$

(where $\theta = h^2/8\pi^2 Ik$). This is also an intractable series, whereas the corresponding integral can be evaluated in closed form.
When $f(J) = (2J+1)e^{-J(J+1)\theta/T}$ the terms up to $(\theta/T)^2$ are

$$f(0) = 1, \quad f'(0) = 2 - \theta/T,$$

$$f^3(0) = -12\theta/T + 12(\theta/T)^2 - \dots,$$

$$f^5(0) = 120(\theta/T)^2 - \dots,$$

and all relevant terms become zero as $J \to \infty$. The integral may be evaluated by making the substitution $u = J(J+1)\theta/T$ to give

$$\int_0^{\infty} (2J+1)e^{-J(J+1)\theta/T} dJ = \frac{T}{\theta}.$$

The Euler–Maclaurin formula (4.22) then gives, to terms in $(\theta/T)^2$,

$$\sum_{J=0}^{\infty} (2J+1)e^{-J(J+1)\theta/T} = \frac{T}{\theta} + \frac{1}{2} - \frac{1}{12}\left(2 - \frac{\theta}{T}\right)$$

$$+ \frac{1}{720}\left(-12\frac{\theta}{T} + 12\left(\frac{\theta}{T}\right)^2\right)$$

$$- \frac{1}{30240}\left(120\left(\frac{\theta}{T}\right)^2\right) + \cdots$$

$$= \frac{T}{\theta}\left(1 + \frac{1}{3}\frac{\theta}{T} + \frac{1}{15}\left(\frac{\theta}{T}\right)^2 + \frac{4}{315}\left(\frac{\theta}{T}\right)^3 + \cdots\right).$$

In this case, the sum may be replaced by the integral when $T \gg \theta$. This applies except for small molecules at low temperatures, for which the series expansion can be used.

CHAPTER 5

Applications of integration

5.1 Plane area

As shown in Section 4.1, the integral

$$\int_a^b f(x)\,dx$$

gives the area between the curve $y = f(x)$, the x-axis and the ordinates $x = a$ and $x = b$.

If, instead, we required the area between the curve and the y-axis, we sum horizontal strips of length x and thickness dy to obtain

$$\int_{f(a)}^{f(b)} x\,dy,$$

where $f(a)$ and $f(b)$ are the values of $y = f(x)$ at $x = a$ and at $x = b$ respectively.

The sum of these two areas must be the difference between the areas of the rectangles $a \times f(a)$ and $b \times f(b)$, as can be seen by reference to Fig. 4.2. This can be proved analytically as follows:

$$\int_a^b f(x)\,dx + \int_{f(a)}^{f(b)} x\,dy$$

$$= \int_a^b f(x)\,dx + \int_a^b x\frac{dy}{dx}\,dx$$

$$= \int_a^b [f(x) + xf'(x)]\,dx$$

$$= \int_{x=a}^{x=b} \mathrm{d}(xf(x))$$

$$= \left[xf(x) \right]_{x=a}^{x=b}$$

$$= bf(b) - af(a).$$

The various manipulations used here should be carefully noted.

These relations give the area between a curve and one of the axes. The area between two curves can be found as the difference between the area under one curve and the area under the other.

The conditions necessary for this interpretation of integration to be valid are that the function should be continuous and single-valued; this means that for any given value of x, the function $y = f(x)$ shall have a single, unique, value. This condition will not be met if the equation contains quadratic or higher-degree terms in y. Thus the equation

$$x^2 + y^2 = r^2$$

represents a circle of radius r, centred at the origin, as shown in Fig. 5.1.

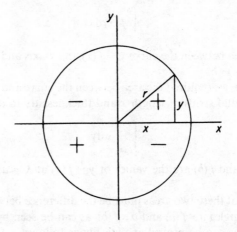

Fig. 5.1

If we simply integrated

$$\int_0^r y \, \mathrm{d}x = \int_0^r \pm \sqrt{(r^2 - x^2)} \, \mathrm{d}x$$

the result would be zero, as the sum of equal positive and negative areas; this is because area has the signs shown in the quadrants of Fig. 5.1.

When a function is not single-valued, we must restrict the range of integration, the area in the first positive quadrant of our circle being obtained by choosing the positive root:

$$\text{area of quadrant} = \int_0^r \sqrt{(r^2 - x^2)}\,dx.$$

This integral may be evaluated by using the substitution $x = r \cos\theta$, $dx = -r \sin\theta\,d\theta$, so as to take advantage of the relation $1 - \cos^2\theta = \sin^2\theta$:

$$\text{area of quadrant} = \int_{\pi/2}^0 r \sin\theta\,(-r \sin\theta\,d\theta)$$

$$= -r^2 \int_{\pi/2}^0 \sin^2\theta\,d\theta$$

$$= \pi r^2/4$$

by using the result of Example 4.3. The area of a circle drawn entirely within the positive quadrant is then πr^2.

5.2   Plane elements of area

In Section 4.1, the use of integration to obtain the area between a curve and the x-axis as

$$\text{area} = \int_a^b y\,dx \tag{5.1}$$

is justified as the sum of areas of vertical strips of height y and width dx. In the same section, integration is interpreted as giving an increment in y by moving along the curve $y = f(x)$, so that

$$y = \int_0^y dy. \tag{5.2}$$

If we now combine equations (5.1) and (5.2) we obtain the area as a double integral

$$\text{area} = \int_a^b \int_0^y dy\,dx. \tag{5.3}$$

The geometrical interpretation of equation (5.3) is that we first obtain the ordinate y by the inner integral and then sum the areas of vertical strips by the outer integral. Alternatively, we can interpret the double integral by regarding $dy\,dx$ as a differential element of area, being that of a rectangle of sides dy and dx. Then

$$dA = dy\,dx$$

and

$$\text{area} = \int dA = \iint dy\,dx.$$

In the same way, we can obtain the area of a circle, using plane polar coordinates, by summing the areas of annular strips of radius r and thickness dr as shown in Fig. 5.2. Given that the circumference of a circle is $2\pi r$ the area of an annular strip becomes, as $dr \to 0$,

$$dA = 2\pi r\,dr.$$

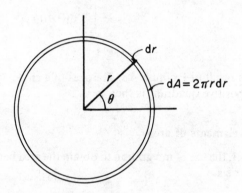

Fig. 5.2

The area of a circle of radius a then becomes

$$\int dA = \int_0^a 2\pi r\,dr = \pi a^2.$$

The differential element of area in plane polar coordinates, shown in Fig. 5.3, is obtained by increasing r by dr and θ by $d\theta$. This requires the element of arc length $ds = r\,d\theta$, which follows from the definition of angle in polar coordinates as angle = (arc length)/radius.
Then

$$dA = r\,d\theta\,dr$$

and

$$\text{area of a circle} = \int_0^a \int_0^{2\pi} r\,d\theta\,dr$$
$$= 2\pi \int_0^a r\,dr = \pi a^2.$$

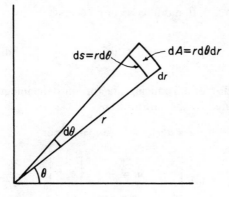

Fig. 5.3

By using polar coordinates we have avoided the difficulty met in Section 5.1 where the area of only a quadrant of the circle could be found. This is because the polar equation of a circle centred at the origin is simply $r = a$, which is a single-valued function.

Example 5.1

An integral that arises frequently in physical applications has the form

$$I = \int_0^\infty e^{-\alpha^2 x^2} dx. \tag{5.4}$$

This may be evaluated by the following device. As a definite integral (i.e. between limits), the variable x is immaterial, so we may rewrite the integral in terms of variable y as

$$I = \int_0^\infty e^{-\alpha^2 y^2} dy. \tag{5.5}$$

We now multiply the two integrals (5.4) and (5.5) together. Since they have common limits, this may be written as

$$I^2 = \int_0^\infty e^{-\alpha^2 x^2} dx \int_0^\infty e^{-\alpha^2 y^2} dy = \int_0^\infty \int_0^\infty e^{-\alpha^2 (x^2 + y^2)} dx\, dy.$$

This double integral is over the positive quadrant of the (x, y) plane. We now transform to polar coordinates, using $x^2 + y^2 = r^2$; the element of area $dx\, dy$ becoming $r\, d\theta\, dr$. The limits for the same positive quadrant become 0 to $\pi/2$ in θ and 0 to ∞ in r, so that

$$I^2 = \int_0^\infty \int_0^{\pi/2} e^{-\alpha^2 r^2} r\,d\theta\,dr = \int_0^\infty \frac{\pi}{2} e^{-\alpha^2 r^2} r\,dr.$$

This has the effect of introducing a factor r into the integral, which can now be evaluated by the substitution

$$u = \alpha^2 r^2, \qquad du = 2\alpha^2 r\,dr.$$

Hence

$$I^2 = \frac{\pi}{4\alpha^2} \int_0^\infty e^{-u}\,du = -\frac{\pi}{4\alpha^2}\left[e^{-u}\right]_0^\infty = \frac{\pi}{4\alpha^2},$$

so that

$$I = \int_0^\infty e^{-\alpha^2 x^2}\,dx = \frac{\sqrt{\pi}}{2\alpha}. \tag{5.6}$$

Example 5.2
The previous example can be used in the integration of similar forms, such as

$$I_1 = \int_0^\infty x^2 e^{-\alpha^2 x^2}\,dx.$$

Integrating by parts, choosing x as the first part, gives

$$I_1 = \left[-\frac{x}{2\alpha^2} e^{-\alpha^2 x^2}\right]_0^\infty + I_2$$

where

$$I_2 = \frac{1}{2\alpha^2} \int_0^\infty e^{-\alpha^2 x^2}\,dx.$$

The first term in I_1 is zero when $x = 0$, but when $x = \infty$ we need to use L'Hôpital's rule (Section 2.10)

$$\lim_{x \to \infty} \frac{x}{e^{\alpha^2 x^2}} = \lim_{x \to \infty} \frac{1}{2\alpha^2 x e^{\alpha^2 x^2}} = 0,$$

so that

$$I_1 = I_2 = \frac{1}{2\alpha^2} \frac{\sqrt{\pi}}{2\alpha} = \frac{\sqrt{\pi}}{4\alpha^3}.$$

5.3   Elements of volume; polar coordinates in three dimensions

In rectangular coordinates, the differential element of volume is that of a cube of sides dx, dy and dz, so that

$$\text{volume} = \int dV = \iiint dx\, dy\, dz. \qquad (5.7)$$

Polar coordinates in three dimensions may be defined in terms of either one angle and two lengths (cylindrical polar coordinates, Fig. 5.4) or two angles and one length (spherical polar coordinates, Fig. 5.5).

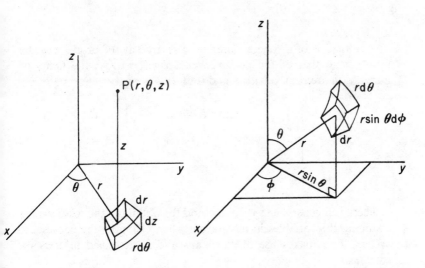

Fig. 5.4   **Fig. 5.5**

Cylindrical polar coordinates	Spherical polar coordinates
$x = r \cos \theta$	$x = r \sin \theta \cos \phi$
$y = r \sin \theta$	$y = r \sin \theta \sin \phi$
$z = z$	$z = r \cos \theta$
$dV = r\, d\theta\, dr\, dz$	$dV = r^2 \sin \theta\, d\theta\, d\phi\, dr$

The corresponding elements of volume are constructed by increasing r by dr, θ by dθ and ϕ by dϕ. When θ is increased by dθ the point P at radius vector r moves in an arc of length rdθ. The volume element in Fig. 5.4 is therefore $r\, d\theta\, dr\, dz$. When ϕ is increased by dϕ in Fig. 5.5, the radius to be considered is $r \sin \theta$, producing an arc of length $r \sin \theta\, d\phi$. The volume element in Fig. 5.5 is therefore $r \sin \theta\, d\phi \;\; r\, d\theta\, dr = r^2 \sin \theta\, d\theta\, d\phi\, dr$. In each case volume will be positive above the (x, y) plane and negative below.

Example 5.3
The volume of a cylinder of radius a and height h with base centred at the origin is

$$V = \int_0^h \int_0^a \int_0^{2\pi} r \, d\theta \, dr \, dz = \int_0^h \int_0^a 2\pi r \, dr \, dz$$

$$= \int_0^h \pi a^2 \, dz = \pi a^2 h.$$

Example 5.4
The volume V of a sphere of radius a centred at the origin must be found from that of the positive hemisphere, so that the limits of integration are 0 to $\pi/2$ for θ and 0 to 2π for ϕ:

$$\tfrac{1}{2} V = \int_0^a \int_0^{2\pi} \int_0^{\pi/2} r^2 \sin \theta \, d\theta \, d\phi \, dr = \int_0^a \int_0^{2\pi} r^2 \, d\phi \, dr$$

$$= \int_0^a 2\pi r^2 \, dr = \tfrac{2}{3} \pi r^3.$$

$$V = 4\pi r^3 / 3.$$

There is a general analytical method for obtaining the relations for changing the variables in integration when changing the coordinate system. This makes use of Jacobians and is described in texts on calculus.

5.4 Line integrals

It is shown in Section 3.5 that, if z is a unique function of x and y, the expression $dz = X \, dx + Y \, dy$ satisfies the cross-differentiation test and is an exact differential. The geometrical interpretation of this is that we can then plot a three-dimensional graph of z against x and y and obtain a unique surface. A physical application of this is when z is a thermodynamic function of the state defined by x and y. Change in z then depends only on the initial and final states, being independent of path.

We may write the change in z as the integral of dz between the endpoints. In two dimensions, with one independent variable, this gives equation (4.7). In three dimensions we obtain

$$\int dz = \int (X \, dx + Y \, dy). \tag{5.8}$$

The right-hand side of (5.8) contains both the variables x and y, so can be regarded as an integral in the (x, y) plane. Between given endpoints the left-hand side gives the change in z and the right-hand side is called a line integral. If we suppose that we move from one value of z to another across the surface in three dimensions along a particular path we can project this path into the (x, y) plane and so define a curve. The line integral is then evaluated along that curve in the (x, y) plane.

To evaluate an integral containing two variables x and y we usually need to know an equation connecting them so as to be able to express the integral in terms of a single variable. This corresponds to choosing a particular path. However, if $X \, dx + Y \, dy$ is an exact differential it can be written as single derivative dz and so integrated. The line integral is then independent of path, the change in z depending only on the endpoints. This also means that the line integral of an exact differential over a closed path is zero because the endpoints then coincide. Thus we may use the condition that the line integral is independent of path as an alternative definition or test for a thermodynamic function of state.

A line integral can be defined and used even when it does not contain an exact differential, and it need not contain both terms on the right-hand side of (5.8). The general characteristic of a line integral is that it contains two variables and so, if not an exact differential, can be evaluated only for a known equation connecting the variables, which means along a particular curve. An example is when the variable of integration is the distance along a curve, which clearly depends on the path taken between the endpoints, and this is discussed in the next section.

5.5   Curve length by integration

When a point moves along a curve, we denote distance travelled by s. Then for increments dx in x and dy in y we can apply Pythagoras's theorem to obtain

$$ds^2 = dx^2 + dy^2.$$

The length of curve between points (x_1, y_1) and (x_2, y_2) is then

$$s = \int_1^2 ds$$

where

$$ds = \sqrt{(dx^2 + dy^2)}$$
$$= \sqrt{[1 + (dy/dx)^2]} \, dx,$$

so that

$$s = \int_1^2 ds = \int_{x_1}^{x_2} \sqrt{[1 + (dy/dx)^2]}\, dx.$$

This integral cannot be evaluated until we know dy/dx in terms of x only, for which we need to know the equation connecting x and y.

Example 5.5
The circumference of a circle of radius r centred at the origin is found from that of the positive quadrant. The equation of the circle is

$$x^2 + y^2 = r^2$$
$$2x\, dx + 2y\, dy = 0$$
$$\frac{dy}{dx} = -\frac{x}{y} = -\frac{x}{\sqrt{(r^2 - x^2)}}.$$

The length of arc is therefore

$$s = \int_0^r \left(1 + \frac{x^2}{r^2 - x^2}\right) dx = \int_0^r \frac{r}{\sqrt{(r^2 - x^2)}}\, dx.$$

Put $x = r\cos\theta$, $dx = -r\sin\theta\, d\theta$,

$$s = \int_0^r \frac{1}{\sqrt{[1 - (x/r)^2]}}\, dx = \int_{\pi/2}^0 \frac{-r\sin\theta}{\sin\theta}\, d\theta = r\int_0^{\pi/2} d\theta = \frac{\pi}{2} r.$$

Hence the circumference of a complete circle is

$$s = 2\pi r.$$

5.6 Applications of multiple integration

Multiple integration is discussed from the analytical point of view in Section 4.4 and in geometrical terms in Sections 5.2 and 5.3. It is generally useful in physical applications, in terms of physical variables of various kinds. This is because differentials can be combined to great advantage in many applications, and finite solutions are then obtained by multiple integration. This is illustrated in the following.

5.6.1 The pressure of a perfect gas

This is given by the kinetic theory of gases in the form of the well known equation

$$pV = \tfrac{1}{3} N m \overline{u^2}.$$

It is derived in standard texts in ways that avoid mathematical complexities such as multiple integration, but this is at the cost of physical clarity. An alternative derivation is presented here in a form which is extended in the next section to include the properties of real fluids.

Since pressure is force per unit area, and force is rate of change of momentum, we can calculate the pressure exerted by collision of gas molecules on unit area of wall from the rate of change of the component of momentum perpendicular to the wall.

In a perfect gas, molecules of mass m are assumed to move freely and independently of each other. Any particular molecule will be in a random position and moving in a random direction with a particular velocity.

Random position and random direction lead to simple calculation. For N molecules in volume V, the density is N/V and the number of molecules in an element of volume dV is the density times that volume, or $(N/V)dV$. Random direction in space can be dealt with using solid angle, this being defined as area/(radius)^2 (Section 1.1), and the total solid angle for a sphere is 4π. The fraction of molecules that will be moving towards an element of surface $d\sigma$ on a sphere of radius r is $d\sigma/4\pi r^2$.

The speed of a particular molecule is a more complicated problem because some molecules move quickly, some slowly, and there is a most probable speed at a given temperature. We have a continuous distribution of velocity amongst the molecules and this is expressed as a velocity distribution $v(u)$; the fraction of molecules with speed between u and $u + du$ depends upon u and is written as $v(u)\,du$.

We multiply together the above three factors to obtain the number of molecules in a volume element dV which are moving towards an element of area $d\sigma$ on a sphere of radius r with a speed between u and $u + du$ as

$$dN = \frac{N}{V}dV\frac{d\sigma}{4\pi r^2}v(u)\,du. \tag{5.9}$$

In Fig. 5.6 we consider molecules moving from an element of volume at P in the gas to collide with an element of area dA on the surface of a wall. Using spherical polar coordinates, with angle ϕ in the plane perpendicular to the paper (compare Fig. 5.5), the element of area dA can be resolved into an element perpendicular to the radius vector **r** as

$$d\sigma = dA \sin\theta \sin\phi. \tag{5.10}$$

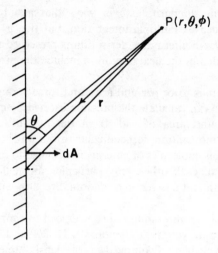

Fig. 5.6

The volume element at P is

$$dV = r^2 \sin\theta \, d\theta \, d\phi \, dr. \tag{5.11}$$

The pressure of the gas is the force exerted per unit area on the wall by elastic collisions of the molecules on the wall; it is the rate of change of momentum perpendicular to the wall on unit area. When a molecule from P collides elastically with the wall, its component of momentum perpendicular to the wall becomes reversed in sign. The component of velocity u perpendicular to the wall is $u \sin\theta \sin\phi$, and the corresponding change in momentum is $2mu \sin\theta \sin\phi$.

The factor that remains to be considered is the number of collisions in unit time; a molecule from P will reach the wall in unit time if the distance r is less than or equal to u.

The total pressure is now found by integration. The differential expression for the pressure, dp, is obtained by multiplying the number of collisions, from equations (5.9), (5.10) and (5.11), by the change in momentum per collision:

$$dp = \frac{N}{V} r^2 \sin\theta \, d\theta \, d\phi \, dr \, \frac{dA \sin\theta \sin\phi}{4\pi r^2} \, v(u) du \, 2mu \sin\theta \sin\phi$$

$$= \frac{N}{V} \frac{m}{2\pi} \sin^3\theta \, \sin^2\phi \, uv(u) \, d\theta \, d\phi \, dr \, dA \, du.$$

The differential variables show this to be a five-fold integral, and we now consider the ranges of integration. We obtain unit area by integrating A from 0 to 1, and unit time by integrating r from 0 to u. Integration of θ and ϕ from 0 to $\pi/2$ ensures that only single values of $\sin\theta$ and of $\sin\phi$ are used and we introduce a factor of 4 so that point P moves through a complete hemisphere based on the wall. The velocity u is integrated from 0 to ∞. Using the notation of Section 4.4 we write

$$p = 4\frac{N}{V}\frac{m}{2\pi}\int_0^\infty uv(u)\,du \int_0^{\pi/2}\sin^3\theta\,d\theta \int_0^{\pi/2}\sin^2\phi\,d\phi \int_0^u dr \int_0^1 dA.$$

It is important here to evaluate the integrals from right to left because the variable u appears in the limits of the integration with respect to r. We have

$$\int_0^1 dA = 1,$$

$$\int_0^u dr = u,$$

$$\int_0^{\pi/2}\sin^2\phi\,d\phi = \pi/4,$$

$$\int_0^{\pi/2}\sin^3\theta\,d\theta = 2/3,$$

so that

$$p = \frac{1}{3}\frac{N}{V}m\int_0^\infty u^2 v(u)\,du.$$

The remaining integral is the average value of the square of the velocity, which we write as $\overline{u^2}$, so that

$$pV = \tfrac{1}{3}Nm\overline{u^2}.$$

5.6.2 Interactions in a real fluid

In a real fluid the molecules exert attractive and repulsive forces on each other and, except at high temperature, the attractive forces are dominant, so that the pressure exerted on the walls is less than it would be for a perfect gas. We denote by $w(r)$ the potential energy of interaction of a pair of molecules separated by a distance r in the bulk fluid; the attractive force is then $-dw(r)/dr$. The number of pairs of molecules at separation r depends upon this potential and we denote by $g(r)$ the pair distribution function, which is the ratio of the number of pairs at separation r to the random number.

Fig. 5.7 shows a plane Oz drawn within bulk fluid. Molecules on the left-hand side interact with those on the right-hand side; by restricting the interaction to a window of unit area we can calculate the total force per unit area, perpendicular to the plane, exerted by the molecules on one side on the molecules on the other.

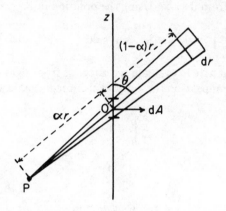

Fig. 5.7

We use point P and unit area at O to define an element of volume on the right-hand side within which molecules would be visible from P. An element of area dA at 0 gives a component d$A \sin \theta \sin \phi$ perpendicular to the radius vector **r**. We define fraction α of the radius vector as being on the left-hand side of the plane, so that the solid angle subtended at P becomes d$A \sin \theta \sin \phi/(\alpha r)^2$. This gives an area visible at radius r of r^2d$A \sin \theta \sin \phi/(\alpha^2 r^2)$. The volume element d$V_R$ in the range r to $r + dr$ on the right-hand side is then

$$dV_R = dA \sin \theta \sin \phi \, dr/\alpha^2. \qquad (5.12)$$

Point P is defined by spherical polar coordinates $(-\alpha r, \theta, \phi)$ and we construct a volume element dV_L at P by using α as the variable at constant r. Then

$$dV_L = (\alpha r)^2 \sin \theta \, d\theta \, d\phi \, d(\alpha r)$$
$$= \alpha^2 r^3 \sin \theta \, d\theta \, d\phi \, d\alpha. \qquad (5.13)$$

The number of pairs of molecules at this separation r is $g(r)$ times the random number; the random number of molecules in dV_R is (N/V)dV_R and in dV_L is (N/V)dV_L. Since each molecule in dV_L interacts with all molecules in dV_R, the number of pairs is

$$\left(\frac{N}{V}\right)^2 dV_L dV_R g(r).$$

The component of the attractive force $-dw(r)/dr$ perpendicular to the plane is $-(dw(r)/dr)\sin\theta\sin\phi$. The differential contribution to the pressure from the intermolecular forces, dp_{int}, is then

$$dp_{int} =$$
$$-\frac{dw(r)}{dr}\sin\theta\sin\phi\left(\frac{N}{V}\right)^2 \alpha^2 r^3 \sin\theta\,d\theta\,d\phi\,d\alpha\,dA\sin\theta\sin\phi\,\frac{dr}{\alpha^2}g(r)$$

$$= -\left(\frac{N}{V}\right)^2 \sin^3\theta\sin^2\phi\, r^3 g(r)\frac{dw(r)}{dr}dr\,d\theta\,d\phi\,d\alpha\,dA.$$

We again have a five-fold integral and the limits are 0 to 1 in A for unit area, 0 to 1 in α to include all pairs for a given direction of the radius vector $\mathbf{r}$, 0 to $\pi/2$ in θ and ϕ, with a compensating factor of 4 to cover a complete hemisphere on each side of the plane, and 0 to ∞ in r:

$$p_{int} =$$
$$-4\left(\frac{N}{V}\right)^2 \int_0^\infty r^3 g(r)\frac{dw(r)}{dr}dr\int_0^{\pi/2}\sin^3\theta\,d\theta\int_0^{\pi/2}\sin^2\phi\,d\phi\int_0^1 d\alpha\int_0^1 dA,$$

$$= -\tfrac{2}{3}\pi\left(\frac{N}{V}\right)^2 \int_0^\infty r^3 g(r)\frac{dw(r)}{dr}dr,$$

which may also be written as

$$p_{int} = -\frac{1}{6}\left(\frac{N}{V}\right)^2 \int_0^\infty g(r)r\frac{dw(r)}{dr}4\pi r^2\,dr.$$

This is an important equation in the theory of liquids and gases.

5.7 The calculus of variations

The determination of stationary (maximum and minimum) values by differentiation of a known function is described in Section 2.7, and this is extended to simultaneous equations in Section 2.8. Another class of problems involving stationary conditions is when the function itself is unknown. For example, we can show that the shortest distance between two points in a plane is a straight line, and that between two points on a sphere is by the great circle route, by finding the function which describes the path of minimum length. We have seen in Section 5.6 that curve length in two dimensions is given by the integral

$$s = \int_{x_1}^{x_2} \sqrt{[1 + (dy/dx)^2]} \, dx. \tag{5.14}$$

Here, as in other problems of this kind, we have a known integral which depends upon an unknown function $y = \phi(x)$. The integrand is a known function of the unknown function y, and can be written in general terms as

$$I = \int_a^b f(x, y, y') \, dx \tag{5.15}$$

where, again, f is known but y is not. We have introduced in (5.15) the device of regarding f as dependent not just on the two variables x and y but also on the third superfluous variable $y' = dy/dx$. As shown in Section 3.4, we may still use the expression for the total differential in the presence of such a superfluous variable, which we shall later do.

The problem now takes the form of finding the function $y = \phi(x)$ which makes the integral I stationary. For this we need to differentiate I with respect to some parameter that produces variation in the function y and so in the integral. This is achieved by the device of writing the unknown function as the sum of two terms

$$y = \phi(x) + \varepsilon \eta(x), \tag{5.16}$$

where $\eta(x)$ is yet another function defined as arbitrary but for the condition that $\eta(x) = 0$ at the endpoints $x = a$ and $x = b$, and ε is an adjustable parameter. If we now vary ε for given $\phi(x)$ and $\eta(x)$ we produce a variation in the curve of y against x between fixed endpoints, as in Fig. 5.8. When $\varepsilon = 0$ we obtain the simple form $y = \phi(x)$. Change in ε produces a change in y at constant x for given $\phi(x)$ and $\eta(x)$.

The integrand in (5.15) is then a function of ε and we can use the stationary condition

$$\frac{\partial I}{\partial \varepsilon} = 0.$$

As shown in the previous section we may interchange the order of differentiation and integration so that

$$\frac{\partial I}{\partial \varepsilon} = \int_a^b \frac{\partial f(x, y, y')}{\partial \varepsilon} \, dx = 0. \tag{5.17}$$

Now change in ε at constant x, $\phi(x)$ and $\eta(x)$ produces changes in y and

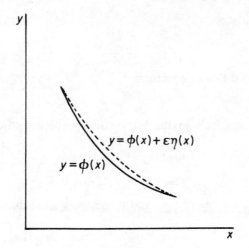

Fig. 5.8

in y' so that

$$\frac{\partial f}{\partial \varepsilon} = \frac{\partial f}{\partial y}\frac{\mathrm{d}y}{\mathrm{d}\varepsilon} + \frac{\partial f}{\partial y'}\frac{\mathrm{d}y'}{\mathrm{d}\varepsilon}$$

and from (5.16)

$$\frac{\mathrm{d}y}{\mathrm{d}\varepsilon} = \eta(x), \qquad \frac{\mathrm{d}y'}{\mathrm{d}\varepsilon} = \frac{\mathrm{d}}{\mathrm{d}\varepsilon}\frac{\mathrm{d}y}{\mathrm{d}x} = \frac{\mathrm{d}}{\mathrm{d}x}\frac{\mathrm{d}y}{\mathrm{d}\varepsilon} = \frac{\mathrm{d}\eta}{\mathrm{d}x} = \eta',$$

so that (5.17) becomes

$$\int_a^b \left(\eta\frac{\partial f}{\partial y} + \eta'\frac{\partial f}{\partial y'} \right)\mathrm{d}x = 0. \tag{5.18}$$

Integration of the second term in (5.18) by parts gives

$$\int_a^b \eta'\frac{\partial f}{\partial y'}\mathrm{d}x = \int_a^b \frac{\partial f}{\partial y'}\eta'\,\mathrm{d}x = \left[\frac{\partial f}{\partial y'}\eta \right]_a^b - \int_a^b \frac{\mathrm{d}}{\mathrm{d}x}\frac{\partial f}{\partial y'}\eta\,\mathrm{d}x. \tag{5.19}$$

By definition, $\eta(x) = 0$ at $x = a$ and at $x = b$ so that the integrated part is zero and (5.18) becomes

$$\int_a^b \eta\left(\frac{\partial f}{\partial y} - \frac{\mathrm{d}}{\mathrm{d}x}\frac{\partial f}{\partial y'} \right)\mathrm{d}x = 0. \tag{5.20}$$

Now equation (5.20) contains $\eta(x)$ which may be any function, so that the only way in which the integral must be zero is when the bracketed part is zero. Hence

$$\frac{\partial f}{\partial y} - \frac{\mathrm{d}}{\mathrm{d}x}\frac{\partial f}{\partial y'} = 0, \tag{5.21}$$

which is called Euler's equation.

Example 5.6
Show that a straight line is the shortest distance between two points in a plane.

When $f(x, y, y')$ is defined by (5.14) we have

$$f(x, y, y') = \sqrt{(1 + y'^2)}$$

and since we are regarding y and y' as separate variables

$$\frac{\partial f}{\partial y} = 0; \qquad \frac{\partial f}{\partial y'} = \tfrac{1}{2}(1 + y'^2)^{-1/2}\, 2y' = y'/\sqrt{(1 + y'^2)},$$

and substitution into (5.21) gives

$$\frac{\mathrm{d}}{\mathrm{d}x}\frac{y'}{\sqrt{(1 + y'^2)}} = 0,$$

or y' is independent of x. This is the equation of a curve of constant slope, which is a straight line.

The application of this technique to three (or more) variables requires only a straightforward extension of the argument. In place of (5.15) an additional variable z will introduce also z' as

$$f = f(x, y, y', z, z').$$

We define two arbitrary functions in place of the one function $\eta(x)$ and two parameters in place of the single parameter ε, one set for y and one set for z. Equation (5.17) becomes the sum of two sets of terms and (5.20) likewise. The condition for the integral in (5.20) to be zero is then that we have an Euler equation of the form of (5.21) in each of the variables x and y.

5.8   Generalized dynamics

The discussion of dynamics in Section 2.11 was based upon Newton's laws of motion. This is not the only basis, however, on which the theory can be built, a useful alternative formulation being due to Hamilton and Lagrange. This regards kinetic and potential energy, rather than force, as fundamental and postulates that a system will move in such a way that the difference between the kinetic energy T and the potential energy V, integrated with respect to time, is stationary. This difference is called the Lagrangian function L and the postulate is called Hamilton's principle, so that

$$I = \int_{t_1}^{t_2} L \, dt = \int_{t_1}^{t_2} (T - V) \, dt \qquad (5.22)$$

is to be stationary, and we can apply the methods of the previous section.

An abstract formulation of this kind is of course more difficult to visualize than Newton's laws because it is not so clearly related to everyday experience. Its great advantage is generality, which arises because it is expressed in terms of the kinetic and potential energies, both of which are scalar quantities. This in turn, means that we are not confined to any particular system of coordinates because we need only to express position in space, not direction as well, which means that we can easily change coordinate systems.

Hamilton's principle is related to the little used concept of action, which is the change in the product of momentum and length. If a field, such as gravity, produces a force F on a particle in the x direction, it moves so that V decreases and T increases by the same amount. The difference in $(T - V)$ is twice the change in potential energy, or twice the work done by the applied force. Denoting momentum by p we have that

$$\text{work} = \int F \, dx \qquad \text{and} \qquad p = \int F \, dt,$$

so that

$$\int L \, dt = 2 \iint F \, dx \, dt = 2 \iint F \, dt \, dx = 2 \int p \, dx,$$

and $\int p \, dx$ is action. An early formulation of quantum theory was in this form, the Wilson and Sommerfeld rule being that $\int p \, dx = nh$ over a closed cycle of the motion. It was shown in Example 1.2 that such a relation is dimensionally correct.

We can use the calculus of variations from the previous section to show that the two formulations of the principles of mechanics are the same. For motion of a particle of mass m in the x direction, the kinetic energy $T = \frac{1}{2}m\dot{x}^2$. When force F acts on the particle in the same x direction the potential energy will change, so that we write that potential energy as $V(x)$. The Lagrangian function L then depends on x

(through V) and on $\dot{x}$ (through T) and we have

$$L(t, x, \dot{x}) = \tfrac{1}{2}m\dot{x}^2 - V(x). \qquad (5.23)$$

Euler's equation (5.21) then becomes

$$\frac{\partial L}{\partial x} - \frac{\mathrm{d}}{\mathrm{d}t}\frac{\partial L}{\partial \dot{x}} = 0, \qquad (5.24)$$

and from (5.23) we have

$$\frac{\partial L}{\partial x} = -\frac{\partial V.}{\partial x} = F,$$

which is the force acting on the particle, and

$$\frac{\partial L}{\partial \dot{x}} = m\dot{x}, \qquad \text{so that} \qquad \frac{\mathrm{d}}{\mathrm{d}t}\frac{\partial L}{\partial \dot{x}} = m\ddot{x},$$

and we obtain $F = m\ddot{x}$, which is Newton's law. Thus Hamilton's principle (5.22) is the same as Newton's law.

The rotation of a diatomic molecule was considered on the basis of Newton's laws in Section 2.11, where it was assumed that masses m_1 and m_2 were rigidly held at a separation r from each other. By using the more powerful technique of equation (5.24) we can consider not only that case but the more general one, where the masses are attracted, or repelled, by each other with a force that is a function of the separation r. The potential energy may be written in the general form $V(r)$; this will apply either to a chemically bonded molecule or to the general case of the interaction between two particles in three-dimensional space.

We use the polar coordinates shown in Fig. 5.9 to define the positions of masses m_1 and m_2 at distances r_1 and r_2 from the centre of mass O of the pair of particles. We suppose the two particles to be moving in space, and so relative to each other, so that both distance r and angle θ change with time. We define the reduced mass m^* of the two particles when they are at separation r as in (2.66) by

$$m^* = \frac{m_1 m_2}{m_1 + m_2},$$

and the rotational kinetic energy from (2.64) and (2.67) becomes in terms of the angular velocity $\dot{\theta} = \omega$

$$\text{rotational KE} = \tfrac{1}{2}m^* r^2 \dot{\theta}^2. \qquad (5.25)$$

Since the distance r between the particles is now a variable, we also have radial kinetic energy given by

$$\text{radial KE} = \tfrac{1}{2}m_1 \dot{r}_1^2 + \tfrac{1}{2}m_2 \dot{r}_2^2. \qquad (5.26)$$

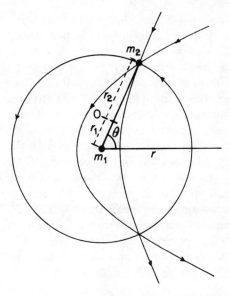

Fig. 5.9

But, from (2.65)

$$r_1 = \frac{rm_2}{m_1 + m_2} \quad \text{and} \quad r_2 = \frac{rm_1}{m_1 + m_2},$$

and (5.26) becomes

$$\text{radial KE} = \tfrac{1}{2}m^*\dot{r}^2. \tag{5.27}$$

Hence

$$L = T - V = \tfrac{1}{2}m^*r^2\dot{\theta}^2 + \tfrac{1}{2}m^*\dot{r}^2 - V(r). \tag{5.28}$$

We now have L in terms of the coordinates r, $\dot{r}$ and θ, $\dot{\theta}$, and we can write Euler's equation (5.24) both in terms of the angular coordinate

$$\frac{\partial L}{\partial \theta} - \frac{\mathrm{d}}{\mathrm{d}t}\frac{\partial L}{\partial \dot{\theta}} = -\frac{\mathrm{d}}{\mathrm{d}t}(m^*r^2\dot{\theta}) = 0. \tag{5.29}$$

and in terms of the radial coordinate

$$\frac{\partial L}{\partial r} - \frac{\mathrm{d}}{\mathrm{d}t}\frac{\partial L}{\partial \dot{r}} = m^*r\dot{\theta}^2 - \frac{\partial V}{\partial r} - \frac{\mathrm{d}}{\mathrm{d}t}(m^*\dot{r}) = 0. \tag{5.30}$$

In equation (5.29), $m^*r^2\dot{\theta} = I\dot{\theta}$ is the angular momentum; this is shown to be constant. In equation (5.30), $m^*r\dot{\theta}^2$ is the centrifugal force in the direction of r opposite to the attractive force between the particles $-\partial V/\partial r$, and $m^*\dot{r}$ is the radial momentum. When the attractive and centrifugal forces balance, $d(m^*r)/dt = 0$ and the radial momentum is constant. In the simplest case of circular motion the radial momentum is zero and r is constant.

The form of the general motion represented by (5.29) and (5.30) depends on the form of $V(r)$. In the important case of an inverse square law attraction, corresponding to gravitational attraction of planetary bodies or the electrostatic interaction between charged particles, the equations represent conic sections (Section 1.7.4). By considering the polar equations of conics, or otherwise, it may be shown that when particles having the same charge approach each other the path of one, relative to the other as focus, is a hyperbola, and for unlike charges we may obtain also a parabola, ellipse or circle for decreasing values of the relative velocity of approach, as illustrated in Fig. 5.9.

This approach to dynamics is called 'generalized' because the equations are the same in any coordinate system. When we denote a generalized position coordinate by q_i we can also define a corresponding, or 'conjugate' 'generalized momentum' by

$$p_i = \frac{\partial L}{\partial \dot{q}_i} = \frac{\partial (T - V)}{\partial \dot{q}_i} = \frac{\partial T}{\partial \dot{q}_i}$$

if V is independent of $\dot{q}_i$.   (5.31)

The significance of this may be seen from the simple case when q_i is an x cartesian coordinate. The momentum is then $p_i = m\dot{x}$ and the kinetic energy T is $\frac{1}{2}m\dot{x}^2$, so that

$$\frac{\partial T}{\partial \dot{q}_i} = \frac{d}{d\dot{x}} (\tfrac{1}{2}m\dot{x}^2) = m\dot{x} = p_i,$$

and if q_i were an angular coordinate θ we would have $T = \frac{1}{2}I\dot{\theta}^2$, and again

$$\frac{\partial T}{\partial \dot{q}_i} = \frac{d}{d\dot{\theta}} (\tfrac{1}{2}I\dot{\theta}^2) = I\dot{\theta} = p_i.$$

The condition in (5.31) that V is independent of $\dot{q}_i$ means that we have a conservative system (Section 3.9.2), or that the work done in any displacement is independent of the velocity (no 'friction').

In these generalized terms we can define also the Hamiltonian function H which is often used in quantum mechanics. The formal definition is

$$H = \sum p_i \dot{q}_i - L.$$ (5.32)

Again for the simple case of motion in the x direction, this gives

$$p_i = m\dot{x}, \qquad \dot{q}_i = \dot{x}, \qquad p_i\dot{q}_i = m\dot{x}^2$$

and
$$L = T - V = \tfrac{1}{2}m\dot{x}^2 - V,$$
so that
$$H = m\dot{x}^2 - (\tfrac{1}{2}m\dot{x}^2 - V)$$
$$= \tfrac{1}{2}m\dot{x}^2 + V$$
$$= T + V = \text{total energy.}$$

This means that the Hamiltonian function has the simple meaning of being the total energy so long as the system is conservative.

We can express Lagrange's equation in terms of coordinate q_i as
$$\frac{\partial L}{\partial q_i} - \frac{\mathrm{d}}{\mathrm{d}t}\frac{\partial L}{\partial \dot{q}_i} = 0,$$
which with (5.31) gives
$$\frac{\partial L}{\partial q_i} = \frac{\mathrm{d}}{\mathrm{d}t}(p_i) = \dot{p}_i.$$

We can then write the equation in terms of H by examining the parameters on which H depends; we have in equation (5.32) that H depends on p_i, $\dot{q}_i$, and on L. In turn L depends on $\dot{q}_i$ (for T) and on q_i (for V). Thus
$$H(p_i, q_i, \dot{q}_i) = \sum p_i \dot{q}_i - L(q_i, \dot{q}_i),$$
so that for the ith coordinate
$$\mathrm{d}H = p_i\,\mathrm{d}\dot{q}_i + \dot{q}_i\,\mathrm{d}p_i - \frac{\partial L}{\partial q_i}\mathrm{d}q_i - \frac{\partial L}{\partial \dot{q}_i}\mathrm{d}\dot{q}_i.$$

But the generalized momentum p_i is defined by equation (5.31) so that the first and last terms cancel to give
$$\mathrm{d}H = \dot{q}_i\,\mathrm{d}p_i - \frac{\partial L}{\partial q_i}\mathrm{d}q_i.$$
Hence
$$\frac{\partial H}{\partial p_i} = \dot{q}_i, \tag{5.33}$$
and
$$\frac{\partial H}{\partial q_i} = -\frac{\partial L}{\partial q_i} = -\dot{p}_i. \tag{5.34}$$

Equations (5.32) and (5.33) are simpler than the equation in terms of the Lagrangian function L in that they have a degree of symmetry; in most problems, however, the Lagrangian function is found first and then the Hamiltonian. The expression (5.33) gives the rate of change of generalized momentum and so may be regarded as a generalized force.

CHAPTER 6

Differential equations

6.1 Significance and notation

Differential equations arise frequently in mathematical and physical theory because the calculus limit so often gives simple relations. In physical theory we may have simple relations for the rates of change with time, distance or temperature of such quantities as energy (thermal conductivity, heat capacity), momentum (force, viscosity) or amount of substance (fluid flow, reaction kinetics, heat of reaction). These theories give rise to equations containing derivatives, and comparison with experiment requires the solution of such equations by integration, the purpose of which is to eliminate the differentials.

The methods of integration discussed in the previous chapter can be used only when the integrand can be expressed in terms of a single variable, so the solution of a differential equation requires the separation of the variables into terms that can be individually integrated. Problems involving only two variables will give rise to total derivatives and ordinary differential equations, whereas if more than two variables are relevant, partial derivatives and partial differential equations will be obtained.

A trivial example of a differential equation is

$$\frac{dy}{dx} = 2x.$$

This may be written with the variables on separate sides of the equation, as

$$dy = 2x \, dx.$$

We may then integrate each side of the equation, to obtain

$$\int dy = \int 2x \, dx$$
$$y = x^2 + C.$$

The integration constant C is important in differential equations.

A simple physical example is when the rate of change of a quantity x with respect to time t is proportional to the quantity itself, so that

$$\frac{dx}{dt} = -kx,$$

giving

$$\int \frac{dx}{x} = -\int k \, dt,$$
$$\ln x = -kt + C.$$

The value of the integration constant can be determined by the use of boundary conditions, such as knowing the initial value x_0 of x at $t = 0$. Then by putting $x = x_0$ and $t = 0$,

$$\ln x_0 = C,$$

and

$$\ln \left(\frac{x}{x_0} \right) = -kt, \qquad x = x_0 e^{-kt}.$$

An alternative interpretation of this form of relation is in terms of a time constant; if an equilibrium system is disturbed by a small amount Δx it will return to equilibrium at a rate which to a first approximation may be assumed to be proportional to the displacement. Then

$$-\frac{d(\Delta x)}{dt} = \frac{1}{\tau} \Delta x,$$

giving

$$\Delta x = \Delta x_0 e^{-t/\tau},$$

so that after time τ the displacement will have decreased to $1/e = 0.37$ of its initial value. This is called the time constant, or relaxation time, of the system and is seen to be the reciprocal of the first-order rate constant k in reaction kinetics.

Not all differential equations possess an analytical solution, but standard methods are available for particular classes of equation. This classification is based upon the concepts of order, degree, linearity and homogeneity.

The order of a differential equation is that of the highest derivative occurring in it, dy/dx being first order, d^2y/dx^2 being second order, and so on. The degree is the power of this derivative. A linear equation is one containing only the first powers of a variable and its derivatives, with no cross-terms, so that

$$a_0(x)\frac{d^2y}{dx^2} + a_1(x)\frac{dy}{dx} + a_2(x)y = f(x) \qquad (6.1)$$

is a second-order linear equation.

The term 'homogeneous' is used in various ways. A first-order equation is homogeneous if each term contains the variables to the same power (Section 6.2.2). A linear equation, on the other hand, is homogeneous if all terms contain the dependent variable, so that $f(x) = 0$ in (6.1).

The solution of a first-order differential equation contains one arbitrary constant, that of a second-order equation contains two constants, and so on.

In the following sections we consider first-order first-degree equations, classed as separable variables, homogeneous, exact or linear, then higher-order linear equations.

The subject of differential equations, and their range of application, is very wide and provides scope for considerable ingenuity in the methods of solution. This chapter does no more than introduce the techniques and some typical applications.

6.2   Equations of first order, first degree

6.2.1   Separable variables

If a differential equation can be written with all terms containing one variable on the left-hand side and all terms containing the other variable on the right-hand side, the solution is merely integration.

Example 6.1

$$\frac{dy}{dx} + 4x = 1.$$

Multiply both sides by dx, giving

$$dy + 4x\,dx = dx,$$

which may be written as

$$dy = (1 - 4x)dx.$$

Integrate both sides

$$\int dy = \int (1 - 4x)dx$$

$$y = x - 2x^2 + C$$

$$y = x(1 - 2x) + C.$$

The geometrical interpretation of this solution is a family of curves, each of which corresponds to a particular value of the integration constant C. Suppose we require the solution passing through the point $(1, 2)$, then we substitute $x = 1$ and $y = 2$ into the equation so as to eliminate the integration constant, giving

$$2 = -1 + C, \qquad C = 3$$

so that the solution to the differential equation becomes

$$y = x(1 - 2x) + 3.$$

Example 6.2
An equation may sometimes be separated into integrable terms even when the variables are not completely separated. Thus if

$$\frac{dy}{dx} = \frac{2x - y}{x + 1}$$

the above procedure gives

$$(x + 1)dy = (2x - y)dx,$$

$$xdy + dy = 2xdx - ydx.$$

The variables are not separable as such due to the terms xdy and ydx, but the combination $xdy + ydx$ is actually the differential of the product xy, so that

$$dy + d(xy) = 2xdx$$

which may be integrated term by term to give

$$\int dy + \int d(xy) = \int 2xdx,$$

$$y + xy = x^2 + C,$$

so that

$$y = \frac{x^2 + C}{1 + x}.$$

Example 6.3
A chemical reaction is described as 'first order' when the rate of production of product is proportional to the first power of the reactant concentration. This is a different use of the term 'order' from the mathematical definition. For the reaction

$$A \xrightarrow{\ k\ } B$$
$$_{(a-x)} \qquad\qquad _x$$

the rate coefficient k is shown, together with the concentrations of the two species at time t. Then

$$\frac{\mathrm{d}x}{\mathrm{d}t} = k(a-x).$$

This is a separable variables equation, so that

$$\frac{\mathrm{d}x}{(a-x)} = k\,\mathrm{d}t,$$
$$-\ln(a-x) = kt + C,$$

and the integration constant C is found from the boundary condition that $x = 0$ at $t = 0$, giving $C = -\ln a$ and

$$\ln\left(\frac{a}{a-x}\right) = kt,$$
$$x = a(1 - e^{-kt}).$$

6.2.2 First-order homogeneous equations

A first-order equation contains only the first differential such as $\mathrm{d}y/\mathrm{d}x$. An equation will be homogeneous if all terms are of the same degree, the ratio $\mathrm{d}y/\mathrm{d}x$ being zero degree, x and $x\,\mathrm{d}y/\mathrm{d}x$ being both first degree, x^2 and xy both second degree, and so on. The standard method of solution is to replace one variable by introducing a new one that is the ratio of the variables. If the variables are x and y we can eliminate y by the substitution $y = zx$.

Example 6.4

$$x\frac{\mathrm{d}y}{\mathrm{d}x} + 2y = 0.$$

Multiply by $\mathrm{d}x$ to give

$$x\,\mathrm{d}y + 2y\,\mathrm{d}x = 0.$$

We now substitute $y = zx$, so that

$$\mathrm{d}y = z\,\mathrm{d}x + x\,\mathrm{d}z,$$

and

$$x(z\,dx + x\,dz) + 2zx\,dx = 0,$$
$$z\,dx + x\,dz + 2z\,dx = 0,$$
$$3z\,dx + x\,dz = 0.$$

The equation will now be of the separable variables kind, to be solved as before. Thus we collect all terms in dx on one side of the equation and terms in dz on the other, to give

$$3z\,dx = -x\,dz,$$
$$\frac{dx}{x} = -\frac{dz}{3z},$$
$$\ln x = -\tfrac{1}{3}\ln z + C,$$
$$\ln x^3 z = C',$$
$$x^3 z = C'',$$

where C, C' and C'' are constants, which are actually related to each other by $C' = 3C = \ln C''$, but the only point of interest is that a constant has been retained in the correct place. We finally return to the original variables by the substitution $z = y/x$, to give

$$x^2 y = C.$$

The correctness of the solution may be checked by differentiation and substitution into the original equation.

Example 6.5

An equation that is homogeneous in the terms containing the variables but also contains an additive constant can be reduced to a simple homogeneous equation by using a substitution to eliminate the constant. Thus if

$$x\frac{dy}{dx} + y = 3$$

we substitute $z = y - 3$, $dz = dy$, to give

$$x\,dy + (y - 3)dx = 0,$$
$$x\,dz + z\,dx = 0,$$
$$\frac{dx}{x} = -\frac{dz}{z},$$
$$\ln x = -\ln z + C$$
$$\ln zx = C,$$
$$zx = C',$$

so that

$$x(y-3) = C.$$

6.2.3   Exact equations

An exact differential is defined in Section 3.5 as being of the form $X\,dx + Y\,dy$ where a function $z(x, y)$ exists with this as its differential:

$$dz = X\,dx + Y\,dy = \left(\frac{\partial z}{\partial x}\right)_y dx + \left(\frac{\partial z}{\partial y}\right)_x dy,$$

so that

$$X = \left(\frac{\partial z}{\partial x}\right)_y \quad \text{and} \quad Y = \left(\frac{\partial z}{\partial y}\right)_x.$$

The test for such an exact differential is the cross-differentiation identity

$$\left(\frac{\partial X}{\partial y}\right)_x = \left(\frac{\partial Y}{\partial x}\right)_y.$$

If a differential equation either has this form or can be converted into it, direct integration is then possible.

Example 6.6

$$dz = 2(x+y)dx + (2x+1)dy.$$

We first test whether or not this is exact, as in Section 3.5, by seeing if cross-differentiation applies, which means looking for equality of

$$\frac{\partial}{\partial y}(2(x+y)) \quad \text{and} \quad \frac{\partial}{\partial x}(2x+1).$$

In this case both are equal to 2 and so it is an exact equation. It may be possible to see by inspection what the function of x and y is that gives the differential equation, but if not we may proceed as follows.

The solution may be obtained from the equations

$$2(x+y) = \left(\frac{\partial z}{\partial x}\right)_y \quad \text{and} \quad 2x+1 = \left(\frac{\partial z}{\partial y}\right)_x.$$

From the first equation

$$z = \int 2(x+y)dx$$

$$= x^2 + 2xy + C(y),$$

this integration being the inverse of partial differentiation with respect to x so that $C(y)$ is a constant only so far as x is concerned; it may be a function of y. Then

$$\left(\frac{\partial z}{\partial y}\right)_x = \frac{\partial}{\partial y}(x^2 + 2xy + C(y))$$

$$= 2x + \frac{dC(y)}{dy} = 2x + 1$$

so that

$$\frac{dC(y)}{dy} = 1, \qquad C(y) = y + C$$

where C is now a numerical constant. The required solution is therefore

$$z = x^2 + 2xy + y + C.$$

6.2.4 Linear equations of first order

These have the form

$$\frac{dy}{dx} + P(x)y = Q(x) \tag{6.2}$$

where $P(x)$ and $Q(x)$ do not contain y but may be functions of x or constants. Writing the equation in the form

$$dy + P(x)y\,dx = Q(x)\,dx \tag{6.3}$$

the solution is obtained by multiplying both sides by the integrating factor

$$e^{\int P(x)dx}.$$

This has the effect of converting the left-hand side of the equation into an exact differential by making cross-differentiation apply (Section 3.5). We show this by multiplying the left-hand side of (6.3) by a function $f(x)$ to give

$$f(x)dy + P(x)yf(x)dx$$

and cross-differentiation gives

$$\frac{\partial f(x)}{\partial x} = \frac{\partial P(x)yf(x)}{\partial y} = P(x)f(x),$$

so that

$$\int \frac{df(x)}{f(x)} = \int P(x)dx$$

$$\ln f(x) = \int P(x)dx,$$

$$f(x) = e^{\int P(x)dx},$$

omitting any integration constant so as to obtain the simplest solution.

Example 6.7

$$\frac{dy}{dx} + 2xy = 4x.$$

Multiply by $e^{\int 2x dx} = e^{x^2}$ gives

$$e^{x^2} dy + 2xye^{x^2} dx = 4xe^{x^2} dx$$

$$d(ye^{x^2}) = 4xe^{x^2} dx.$$

Integrating both sides

$$ye^{x^2} = 2e^{x^2} + C, \qquad y = 2 + Ce^{-x^2}.$$

Example 6.8

Successive kinetically 'first-order' chemical reactions give linear differential equations. An extension of Example 6.3 is

$$A \xrightarrow{k_1} B \xrightarrow{k_2} C$$
$$_{(a-x)} \qquad _{(x-y)} \qquad _{y}$$

where the rate coefficients k_1 and k_2 are shown, together with the concentrations of the various species at time t. The rate of production of C is given by

$$\frac{dy}{dt} = k_2(x - y).$$

This may be seen to be a first-order linear differential equation by writing it in the form

$$\frac{dy}{dt} + k_2 y = k_2 x.$$

The integrating factor is $e^{\int k_2 dt} = e^{k_2 t}$, giving

$$e^{k_2 t} dy + k_2 ye^{k_2 t} dt = k_2 xe^{k_2 t} dt,$$

$$d(ye^{k_2 t}) = k_2 xe^{k_2 t} dt.$$

The right-hand side contains the variable x, which may be eliminated by applying the result of Example 6.3 to the first stage of the reaction. Thus

$$x = a(1 - e^{-k_1 t}).$$

Substitution and integration then gives

$$y = a - \frac{k_2 a}{k_2 - k_1} e^{-k_1 t} + Ce^{-k_2 t}.$$

The integration constant C may be found from the initial conditions that $y = 0$ at $t = 0$, giving

$$y = a[1 - (k_2 e^{-k_1 t} - k_1 e^{-k_2 t})/(k_2 - k_1)].$$

Fig. 6.1 shows typical curves of $(a - x)$, $(x - y)$ and y, each plotted against time t.

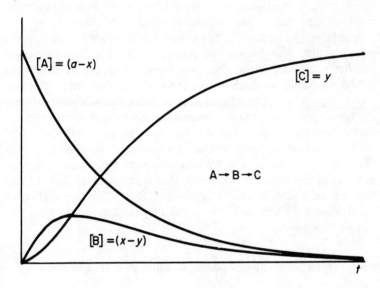

[A] = (a−x)

[C] = y

A → B → C

[B] = (x−y)

t

Fig. 6.1

6.3 Linear differential equations

A linear equation is one containing only the first powers of a variable and its derivatives, the general form of linear equation of order n being

$$a_0(x)\frac{d^n y}{dx^n} + a_1(x)\frac{d^{n-1} y}{dx^{n-1}} + \ldots + a_n(x)y = f(x). \tag{6.4}$$

There is a general method of solution for such an equation only when the coefficients $a_0, a_1, \ldots$ are constants, and this is discussed in Section 6.3.2. An equation in which all terms are functions of the dependent variable y is called homogeneous, which is when $f(x) = 0$ in (6.4), and this is discussed in Section 6.3.1.

An important consequence of an equation being linear is that linear combinations of solutions may also be used. For example, if $y = \alpha(x)$

and $y = \beta(x)$ are both solutions of the homogeneous equation

$$a_0 \frac{d^2 y}{dx^2} + a_1 \frac{dy}{dx} + a_2 y = 0,$$

then $y = \alpha(x) + \beta(x)$ is also a solution. This follows because the equation may then be separated into parts after making the substitution, one part containing α and the other part containing β, and both parts are zero since α and β are both solutions of the equation.

A physical system is called linear if its variables are linearly related, meaning that multiplication of one variable by a factor increases another by the same factor. Thus a linear amplifier is one for which if we double the input we double the output. For a 'first-order' chemical reaction the rate is proportional to concentration and a series of such reactions gives rise to a linear differential equation, as in Example 6.8. The differential equations describing linear physical systems will be linear not only in the dependent variable y and its derivatives but also in the independent variable x. We may use linear differential equations not only to describe physical systems that are truly linear, but also as a first approximation to non-linear systems. This is because the effect of small enough disturbances will be given by the first, linear, term in a Taylor expansion.

6.3.1 Homogeneous linear equations with constant coefficients

These have the form

$$a_0 \frac{d^n y}{dx^n} + a_1 \frac{d^{n-1} y}{dx^{n-1}} + \ldots + a_n y = 0. \tag{6.5}$$

The substitution $y = e^{mx}$ reduces the solution of the equation to that of the auxiliary.algebraic equation in m.

Example 6.9

$$\frac{d^2 y}{dx^2} + \frac{dy}{dx} - 2y = 0.$$

Put

$$y = e^{mx}, \qquad \frac{dy}{dx} = m e^{mx}, \qquad \frac{d^2 y}{dx^2} = m^2 e^{mx},$$

then

$$e^{mx}(m^2 + m - 2) = 0.$$

This equation will be satisfied by solving the auxiliary equation

$$m^2 + m - 2 = 0,$$

which factorizes to give either $m = 1$ or $m = -2$, so that the differential equation is satisfied by either $y = e^x$ or $y = e^{-2x}$. It is also satisfied by $y = C_1 e^x$ and by $y = C_2 e^{-2x}$, where C_1 and C_2 are arbitrary constants. Furthermore, since the equation is linear, we may also write

$$y = C_1 e^x + C_2 e^{-2x}.$$

This last solution contains the two arbitrary constants expected in the solution of a second-order equation and is the most general form, the earlier solutions corresponding to values of zero or unity for the arbitrary constants.

This method gives solutions whenever the auxiliary equation has real roots. If the auxiliary equation were to give equal roots, however, we would obtain only a solution containing one arbitrary constant. In that case, multiplication of that root by $C_2 x$ will be found to produce the required second solution.

A solution to the differential equation may exist even when the roots of the auxiliary equation are imaginary since these appear as imaginary exponents in the solution. From Section 1.10

$$e^{iy} = \cos y + i \sin y$$

and

$$e^{-iy} = \cos y - i \sin y.$$

Now, the arbitrary constants used in the solution of the differential equation may also be complex numbers and these will then produce real solutions, as in the following example.

Example 6.10

$$\frac{d^2 y}{dx^2} + y = 0.$$

Put $y = e^{mx}$, so that

$$e^{mx}(m^2 + 1) = 0$$

and the auxiliary equation becomes

$$m^2 + 1 = 0, \qquad m = \pm i,$$

giving the general solution

$$y = A e^{ix} + B e^{-ix}.$$

Taking A and B to be complex numbers, we put $A = a + ib$ and

$B = a - ib$, giving

$$y = (a + ib)(\cos x + i \sin x) + (a - ib)(\cos x - i \sin x)$$
$$= 2a \cos x - 2b \sin x,$$

or

$$y = C \cos x + D \sin x,$$

which is wholly real.

Example 6.11
An equation of the form of Example 6.10 is obtained for simple harmonic motion, when a particle of mass m experiences a restoring force proportional to its displacement x. Then

$$\text{force} = m \frac{d^2 x}{dt^2} = -kx,$$

$$\frac{d^2 x}{dt^2} + \frac{k}{m} x = 0.$$

Put

$$x = e^{at}, \qquad \frac{d^2 x}{dt^2} = a^2 e^{at},$$

then

$$e^{at}\left(a^2 + \frac{k}{m}\right) = 0, \qquad a = \pm i \sqrt{(k/m)}$$

giving

$$x = A e^{it\sqrt{(k/m)}} + B e^{-it\sqrt{(k/m)}},$$

and when A and B are allowed to be complex,

$$x = C \cos\left[t\sqrt{(k/m)}\right] + D \sin\left[t\sqrt{(k/m)}\right].$$

By using the trigonometric identity

$$\sin(A + B) = \sin A \cos B + \cos A \sin B$$

we can write the solution in the form

$$x = A \sin(\omega t + \phi),$$

where the constants A and ϕ are simply related to the constants C and D, and

$$\omega = \sqrt{(k/m)}.$$

This equation suggests the interpretation in terms of polar coordinates shown in Fig. 6.2; point P moves in a circle of radius A with

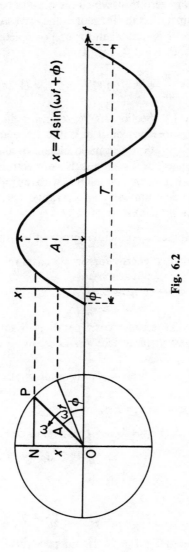

$x = A \sin(\omega t + \phi)$

Fig. 6.2

angular velocity ω, and x is the projection ON of OP onto the vertical axis. Point N moves up and down on the vertical axis with simple harmonic motion. The amplitude of the motion is A, and the angle ϕ gives an arbitrary zero from which the motion begins at time $t = 0$. A graph of x against t has the form of a sine wave as shown in the figure. The time period T is that taken for one complete cycle of the motion and the frequency v is $1/T$, so that

$$T = \frac{2\pi}{\omega} = 2\pi \sqrt{(m/k)}, \qquad v = (1/2\pi)\sqrt{(k/m)}.$$

If OP is a vector (Section 3.9) representing the dipole moment of a rotating polar molecule, or if x is the fluctuating dipole moment accompanying vibration of a molecule, an oscillating electric field is produced which will be the same as that of electromagnetic radiation of the same frequency, so that interaction with the radiation can occur and an absorption spectrum will be observable.

From equation (2.58),

$$\text{force} = -\text{potential gradient} = -\mathrm{d}V/\mathrm{d}x,$$

so that the potential energy for simple harmonic motion is given by

$$V = -\int F\,\mathrm{d}x = \tfrac{1}{2}kx^2,$$

and the curve of V against x is a parabola (Example 1.7).

The velocity at displacement x is given by

$$v = \frac{\mathrm{d}x}{\mathrm{d}t} = \omega A \cos(\omega t + \phi),$$

so that the kinetic energy T becomes

$$\begin{aligned}
T = \tfrac{1}{2}mv^2 &= \tfrac{1}{2}m\omega^2 A^2 \cos^2(\omega t + \phi) \\
&= \tfrac{1}{2}m\omega^2 A^2 [1 - \sin^2(\omega t + \phi)] \\
&= \tfrac{1}{2}m\omega^2 (A^2 - x^2).
\end{aligned}$$

But $\omega^2 = k/m$, so that

$$T = \tfrac{1}{2}kA^2 - \tfrac{1}{2}kx^2$$

giving total energy $T + V = \tfrac{1}{2}kA^2$, independent of x. Thus when $x = 0$ the energy is wholly kinetic, when $x = A$ it is wholly potential, and we have an interchange between kinetic and potential energies during the motion.

The vibration of a diatomic molecule occurs by extension and compression of a chemical bond and we can, to a first approximation, assume that the restoring force is proportional to the displacement. Given that we resolve the general motion of a molecule into motion of the centre of mass and motion relative to the centre of mass, the motion will be as shown in Fig. 6.3, each atom vibrating relative to the centre of mass O. The restoring forces on the two masses are equal and opposite, so that

$$\text{force } F = -m_1 \frac{d^2r_1}{dt^2} = -m_2 \frac{d^2r_2}{dt^2}.$$

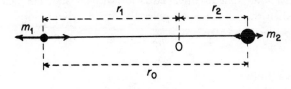

Fig. 6.3

But from equations (2.65) and (2.66)

$$r_1 = \frac{rm_2}{m_1 + m_2} \quad \text{and} \quad m^* = \frac{m_1 m_2}{m_1 + m_2}$$

so that

$$F = -m^* \frac{d^2r}{dt^2}$$

and the motion will be as described above with the usual substitution of reduced mass m^* for m, zero displacement corresponding to the equilibrium separation r_0 so that x becomes $(r - r_0)$.

This simple model of molecular vibration will apply only to small displacements; in vibrationally excited states the restoring force will no longer be proportional to the displacement and the curve of potential energy against r will depart from the parabola corresponding to simple harmonic motion.

Example 6.12
Simple harmonic motion will occur only in the absence of frictional losses. The motion of real macroscopic systems will be subject to resistance which, often to a good first approximation, is proportional to the velocity. This applies both to the vibration of an object immersed in a fluid such as air or a liquid, and to electrical oscillations in the

presence of electrical resistance. The process then becomes irreversible and potential and kinetic energy are gradually converted into heat, so that the vibration decays at a rate depending on the resistance. This is called damping. The restoring force is then given by

$$\text{force} = -kx - \beta \frac{dx}{dt} = m \frac{d^2x}{dt^2},$$

where the frictional, or damping, force is β times the velocity and acts in the same direction as the restoring force when dx/dt is positive and so the displacement increasing.

This is a homogeneous linear equation with constant coefficients and the usual substitution $x = e^{at}$ gives the auxiliary equation

$$a^2 + \frac{\beta a}{m} + \frac{k}{m} = 0,$$

with the solution

$$a = -\frac{1}{2m} [\beta \pm (\beta^2 - 4mk)].$$

The behaviour then depends on the relative magnitudes of β^2 and $4mk$, which means on the extent of the damping.

When heavily damped, $\beta^2 > 4mk$ and the auxiliary equation has real roots, say $-a_1$ and $-a_2$; the solution then has the form of Example 6.9,

$$x = C_1 e^{-a_1 t} + C_2 e^{-a_2 t},$$

which corresponds to exponential decay of any imposed displacement. This is illustrated as curve A in Fig. 6.4.

When $\beta^2 = 4mk$, the auxiliary equation has equal real roots, giving the solution

$$x = C_1 e^{-(\beta/2m)t} + C_2 t e^{-(\beta/2m)t}.$$

This also gives exponential decay, the damping force being just sufficient to prevent oscillation, but allowing the decay to occur most quickly (curve B in Fig. 6.4). This is called the critical damping condition.

If β were zero we would have purely imaginary roots to the auxiliary equation as in Example 6.11, and so simple harmonic motion. When $0 < \beta^2 < 4mk$, less than the critical amount of damping is present and the roots of the auxiliary equation will be complex, of the form

$$a = d \pm if,$$

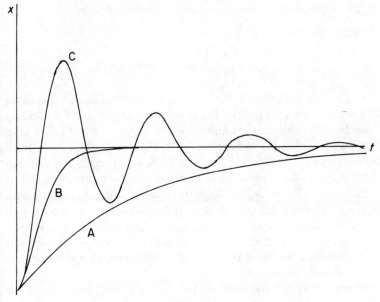

Fig. 6.4

where d (for decay) and f (for frequency) are real numbers given by

$$d = -\beta/(2m) \quad \text{and} \quad f = [\sqrt{(4mk - \beta^2)}]/2m$$

$$= \sqrt{\left(\frac{k}{m} - \frac{\beta^2}{4m^2}\right)}.$$

The solution is then the product of terms arising from the real and the imaginary parts; the real part is exponential decay and the imaginary part gives periodic oscillation of frequency $f/(2\pi)$, the result being a damped oscillation of form C in Fig. 6.4. We notice that the frequency of the damped oscillation is lower than the value $(1/2\pi)\sqrt{(k/m)}$ which would be obtained in the absence of damping.

The critical damping condition is very important in practice because that gives the fastest response without oscillation. This is achieved by adjusting the value of β: in a simple moving-coil galvanometer system by adjusting the circuit resistance to equal that for critical damping, and in more complicated control systems by adjusting the control signal input which depends upon the derivative dx/dt.

6.3.2   *General linear equation with constant coefficients*

This has the form

$$a_0 \frac{d^n y}{dx^n} + a_1 \frac{d^{n-1} y}{dx^{n-1}} + \ldots + a_n y = f(x). \qquad (6.6)$$

We again use the principle that the n adjustable constants expected in the solution of an nth-order linear equation are obtained by adding together the n separate solutions, as in Example 6.9. In this case, however, we also take advantage of the fact that only one of these solutions need give the right-hand side of the equation; the rest can then be solutions of the corresponding homogeneous equation, which is with $f(x) = 0$. To distinguish these two kinds of solution, the ones with $f(x) = 0$ are called the complementary function and the solution with the given value of $f(x)$ is called the particular integral. The general solution of the equation is then the sum of these two. We will see that this separation of solutions has not only mathematical significance but physical significance as well.

There are systematic methods for finding a particular integral for equations of various types, details of which are given in texts on differential equations. In simple cases, however, it is often possible to find a solution by trial; in particular, when $f(x)$ is a polynomial, or an exponential or trigonometric function, the same function with arbitrary coefficients should be used as a trial solution. This is substituted into the equation and, if successful, the values of the coefficients will be found; if not, we try multiplying the trial solution by x.

Example 6.13

$$\frac{d^2 y}{dx^2} + \frac{dy}{dx} - 2y = 1 - 2x.$$

Here $f(x) = 1 - 2x$, which is a first-degree polynomial, so that we try the general first-degree polynomial $y = ax + b$ as a solution. Substituting this into the equation gives

$$a - 2(ax + b) = 1 - 2x,$$

which is satisfied by $a = 1$ and $b = 0$. A particular integral is therefore $y = x$. The complementary function for this equation was found in Example 6.9 to be

$$y = C_1 e^x + C_2 e^{-2x},$$

and the general solution is the sum of the complementary function and a particular integral, or

$$y = C_1 e^x + C_2 e^{-2x} + x.$$

Example 6.14

Show that the method of this section gives the same solution as the method of Section 6.2.4 by solving the equation

$$\frac{dy}{dx} + y = 3x.$$

We are here comparing the method that applies to a linear equation of any degree, so long as the coefficients are constants, with the method that applies to any linear equation of the second degree.

We write the linear homogeneous form of the equation,

$$\frac{dy}{dx} + y = 0,$$

and put $y = e^{mx}$,

$$e^{mx}(m+1) = 0, \qquad m = -1$$

to yield the complementary function

$$y = Ce^{-x}.$$

The right-hand side of the original equation is again a polynomial of first degree, so we try $y = ax + b$ to give

$$\frac{dy}{dx} + y = a + ax + b$$

$$= 3x \qquad \text{if } a = 3, b = -3,$$

giving $y = 3x - 3$ as a particular integral, so that the general solution becomes

$$y = Ce^{-x} + 3x - 3.$$

The solution corresponding to Section 6.2.4 is to multiply the equation by $e^{\int P dx} = e^{\int dx} = e^x$, and the equation becomes

$$e^x \frac{dy}{dx} + ye^x = 3xe^x,$$

$$d(ye^x) = 3xe^x dx,$$

$$ye^x = 3e^x(x-1) + C$$

so that

$$y = 3x - 3 + Ce^{-x}$$

as before. The same equation is solved by a third method in Example 6.16.

Example 6.15
We have discussed damped harmonic motion in Example 6.12 and have interpreted the damping as a dissipation of kinetic and potential energy as heat. A common physical situation is when such oscillation is sustained by a forcing function in the form of a periodic input of energy which may be applied mechanically, or as an applied alternating voltage to an electrical circuit, or as incident radiation of a particular frequency on a molecular system. In order to sustain oscillation the forcing function must be 'in tune' with the oscillation; when we apply a sinusoidal forcing function $-F \sin pt$ the equation becomes

$$\text{force} = -kx - \beta \frac{dx}{dt} - m \frac{d^2x}{dt^2} = -F \sin pt.$$

The complementary function will be as in Example 6.12. A particular integral is sought by using a trial function of the same form as the right-hand side of the equation, so we put

$$x = A \cos pt + B \sin pt,$$

$$\frac{dx}{dt} = -Ap \sin pt + Bp \cos pt,$$

$$\frac{d^2x}{dt^2} = -Ap^2 \cos pt - Bp^2 \sin pt$$

$$= -p^2 x,$$

and the equation becomes

$$-mp^2 x - \beta Ap \sin pt + \beta Bp \cos pt + kx = F \sin pt,$$

which is satisfied if

$$k = mp^2, \qquad p = \sqrt{(k/m)},$$
$$\beta Ap = -F, \qquad A = -F/(\beta p) = -(F/\beta)\sqrt{(m/k)},$$
$$B = 0.$$

The particular integral is then

$$x = -\frac{F}{\beta} \sqrt{\left(\frac{m}{k}\right)} \cos\left[t \sqrt{\left(\frac{k}{m}\right)}\right].$$

The damped oscillation conditions are, in the notation of Example

6.12, $0 < \beta^2 < 4mk$ and $a = d \pm if$, so that the complementary function is (compare Example 6.10).

$$x = e^{-\beta t/(2m)}(C_1 \cos ft + C_2 \sin ft),$$

and the general solution is

$$x = e^{-\beta t/(2m)}(C_1 \cos ft + C_2 \sin ft) - \frac{F}{\beta} \sqrt{\left(\frac{m}{k}\right)} \cos\left[t \sqrt{\left(\frac{k}{m}\right)}\right].$$

The physical significance of this is that the motion is the damped harmonic motion that would follow a displacement in the absence of the forcing function, together with the final term in the equation. The condition that was chosen to obtain the particular integral was when $p = \sqrt{(k/m)}$, which is the value of ω for simple harmonic motion (Example 6.11). This corresponds to applying a forcing function at the frequency $\omega/2\pi$ which the system would display if it were undamped, in spite of the fact that the real damped oscillation frequency is lower.

If the system were displaced and then the forcing function applied, it would display damped motion for a time until the contribution from the first term (the complementary function) in the general solution had decayed, followed by sustained oscillation given by the second term only. The condition when the forcing frequency is equal to the natural undamped frequency is called resonance; this is the completely contrary condition to critical damping.

6.3.3 Linear equations of second order

The general second-order linear equation can be reduced to a first-order linear equation if a solution can be found by inspection. Given the equation

$$\frac{d^2y}{dx^2} + p(x)\frac{dy}{dx} + q(x)y = f(x) \tag{6.7}$$

if $y = y_1$ is a solution to the corresponding homogeneous equation with $f(x) = 0$, then put $y = y_1 u(x)$, so that

$$\frac{dy}{dx} = y_1 \frac{du}{dx} + u\frac{dy_1}{dx},$$

$$\frac{d^2y}{dx^2} = y_1 \frac{d^2u}{dx^2} + \frac{dy_1}{dx}\frac{du}{dx} + u\frac{d^2y_1}{dx^2} + \frac{dy_1}{dx}\frac{du}{dx}.$$

Substitution into (6.7), remembering that when $y = y_1$ the expression on the left-hand side of (6.7) is zero, gives

$$\frac{d^2u}{dx^2} + 2\frac{dy_1}{dx}\frac{du}{dx} = \frac{f(x)}{y_1}.$$

This is a first-order linear equation, which may be solved for du/dx, from which $u(x)$ may be found by integration.

6.4 Integral transforms

Substitution methods are often used to transform from one independent variable to another. This may be achieved not only by algebraic substitution but also by the use of substitutions in the form of integrals. We consider two forms of such integral transformation: Laplace transforms, which are useful in solving differential equations; and Fourier transforms, which are used in instrumentation.

6.4.1 Laplace transforms

The Laplace transform $L(f(x))$ of the function $f(x)$ is defined by

$$L(f(x)) = \int_0^\infty e^{-px}f(x)dx. \tag{6.8}$$

The limits of integration with respect to x mean that the variable x disappears after integration; we obtain instead an expression in terms of the new variable p, which is why this is called a transformation. This is useful in solving differential equations for three reasons: the first is that the integral can be evaluated for simple functions and, indeed, is given in tables. The second reason is that it is a linear operator, which means that multiplying by a constant, or adding or subtracting terms in $f(x)$, simply means multiplying, adding or subtracting terms in $L(f)$. The third reason, which allows it to be useful in solving differential equations, is that the Laplace transform of derivatives dy/dx, d^2y/dx^2, ... are also simple algebraic functions of the new variable p. The terms in a linear differential equation can in this way be converted into an algebraic equation in p, which can be solved to give the Laplace transform of y in terms of p. The tables of transforms are then consulted again in the inverse sense of: knowing the transform $L(y)$, find y and so the solution of the equation. Only a simple example will be given to illustrate the method, the reader being referred to a text containing a full table of transforms for more details. This technique is particularly useful for the equations describing electrical circuits.

A particular example of a Laplace transform is the gamma function $\Gamma(x)$, defined by

$$\Gamma(x) = \int_0^\infty t^{x-1}e^{-t}dt, \tag{6.9}$$

so that $\Gamma(x)$ is the transform of t^{x-1}. This may be evaluated by integration by parts. We write the function for $(x+1)$ as

$$\Gamma(x+1) = \int_0^\infty t^x e^{-t}dt$$

$$= -[t^x e^{-t}]_0^\infty + x\int_0^\infty t^{x-1}e^{-t}dt$$

$$= x\Gamma(x). \tag{6.10}$$

Repeated integration by parts then gives

$$\Gamma(x+1) = x!\,\Gamma(1)$$

where

$$\Gamma(1) = \int_0^\infty e^{-t}dt = 1$$

so that

$$\Gamma(x+1) = \int_0^\infty t^x e^{-t}dt = x!. \tag{6.11}$$

If we substitute n for x and px for t in (6.11) we obtain the form (6.8) as

$$p^n\int_0^\infty x^n e^{-px}p\,dx = p^{n+1}L(x^n) = n! \tag{6.12}$$

In particular, when $n = 1$ and when $n = 0$

$$L(x) = 1/p^2 \quad \text{and} \quad L(1) = 1/p. \tag{6.13}$$

The Laplace transform of dy/dx is given by

$$L(dy/dx) = \int_0^\infty e^{-px}\frac{dy}{dx}dx$$

$$= [ye^{-px}]_0^\infty + p\int_0^\infty ye^{-px}dx$$

$$= -y(0) + pL(y), \tag{6.14}$$

where $y(0)$ is the value of y when $x = 0$.

Example 6.16

As a simple example we use Laplace transforms to solve the same equation as Example 6.14

$$\frac{dy}{dx} + y = 3x.$$

Taking Laplace transforms of each term gives

$$-y(0) + pL(y) + L(y) = 3/p^2.$$

Solving for $L(y)$ we obtain

$$L(y) = \frac{1}{p+1} y(0) + \frac{3}{p^2(p+1)}$$

$$= \frac{1}{p+1} y(0) - \frac{3}{p} + \frac{3}{p^2} + \frac{3}{p+1}$$

where we have used partial fractions for the last term. We now take the inverse transform of each term. From a table of transforms, or by evaluating the integral, we find

$$L(e^{-x}) = 1/(p+1)$$

which, together with (6.13), gives

$$y = y_0 e^{-x} - 3 + 3x + 3e^{-x}$$

$$= (y_0 + 3)e^{-x} - 3 + 3x,$$

which is the same solution as obtained by the complementary function and particular integral approach in Example 6.14 when $C = y_0 + 3$.

Example 6.17

The Schrödinger equation in one dimension has the form

$$\frac{\partial^2 \psi}{\partial x^2} = -\frac{8\pi^2 m}{h^2} (E - V)\psi.$$

For free translation of a particle of mass m between walls separated by distance a, the potential energy V does not change and so may be taken as zero, and the wavefunction ψ must be zero at the walls, which are defined by $x = 0$ and $x = a$.

This equation may conveniently be solved by using Laplace transforms. From tables, or by integration similar to (6.14),

$$L(d^2\psi/dx^2) = p^2 L(\psi) - p\psi(0) - \psi'(0).$$

Taking transforms of the equation, with $V = 0$ and $\psi(0) = 0$,

$$p^2 L(\psi) - \psi'(0) = -\frac{8\pi^2 m}{h^2} E L(\psi).$$

Putting $8\pi^2 mE/h^2 = \lambda^2$ we obtain

$$L(\psi) = \psi'(0)/(p^2 + \lambda^2).$$

Again from tables

$$L(\sin \lambda x) = \lambda/(p^2 + \lambda^2)$$

giving the solution to the equation

$$\psi = \frac{\psi'(0)}{\lambda} \sin \lambda x,$$

and at $x = a$, $\psi = 0$ so that

$$\sin \lambda a = 0, \qquad \lambda a = n\pi.$$

Hence

$$E = \frac{\lambda^2 h^2}{8\pi^2 m} = \frac{n^2 h^2}{8ma^2}.$$

6.4.2 *Fourier transforms*

Fourier transforms are of practical importance because their use enables instruments to be designed and used in such a way that results can be obtained very quickly. This is an advantage because measurements of weak signals always have superimposed upon them the effects of random disturbances called noise. Since noise produces signals of random sign these average to zero for a large enough number of repeated measurements, leaving the required signal less obscured by noise.

Sine and cosine functions are periodic, so that a graph of $\sin \omega t$ against t has the form of a wave of frequency $\omega/2\pi$. If we superimpose sine and cosine functions of different frequencies we produce a complicated pattern which, given enough terms, can be adjusted to fit any particular curve as closely as we may wish. The inverse of this, the resolution of a given function into a sum of sine and cosine terms, is Fourier analysis, for which there are standard procedures. In the limit when the number of terms becomes infinite we may replace the sum by an integral; we then obtain the Fourier transforms of $f(\omega)$ as

$$F(t) = \frac{1}{2\pi} \int_{-\infty}^{+\infty} f(\omega) e^{-i\omega t} d\omega$$

and

$$f(\omega) = \int_{-\infty}^{+\infty} F(t)e^{i\omega t}dt$$

which relate functions of frequency (ω) and functions of time (t). The complex exponential function is expressed in sine and cosine terms in Section 1.10.

An appropriate text should be consulted for the derivation and use of Fourier analysis and transforms. Their application in spectroscopy is to transform measurements of the decay with time of the response induced by an irradiating pulse into the required spectrum as a function of frequency. The mathematical and physical principles may be brought together by observing, first, that a pulse is equivalent to a superposition of different frequencies with the major components near the frequency of the pulsed radiation, so that a spectrum of absorption frequencies can all be excited simultaneously and, secondly, that the resulting decay, or relaxation in time, that is observed can be transformed into a superposition of exponential decays in time of a set of frequencies, the set being the required spectrum.

CHAPTER 7

Experimental error and the method of least squares

7.1 Significance

Experimental measurements must be made and used with their probable accuracy in mind. Some properties can and should be measured more precisely than others; for example, it is as absurd to measure the length of a piece of string to 0.01 mm as it would be to weigh a block of gold to the nearest kilogram. The least that should be done is to quote only figures that are believed to be reliable, so that a result given as 2.1 implies that it is between 2.05 and 2.15, and 2.1234 implies that it is between 2.12335 and 2.12345. It is preferable to give an explicit statement of precision, such as 2.123 ± 0.024, the result and the uncertainty both being quoted to the same number of decimal places. This implies that the results have been analysed with great care because two figures are quoted in the uncertainty, which is the most that can ever be achieved. It does not imply that it is not possible to obtain greater precision in the measurement because improved equipment and better instruments may well produce a more precise result; it does imply, however, that such measurements will not give results outside the bounds set on the existing measurement. This chapter is concerned with the methods of analysing experimental data which are necessary to achieve such a high level of confidence in the statement of results.

Experimental errors are of two kinds, systematic and random. A typical systematic error would arise if length were measured using a ruler calibrated at one temperature but used at a higher one. All measurements would then be short by an amount proportional to the length being measured. Systematic errors arise from inadequacy in the calibration of instruments, or from inadequacy in the design of apparatus or in the technique used.

The term 'random' is used in error analysis in a particular sense. Instability, due to poor control, mechanical vibration or thermal or electrical noise, produces a scatter in the experimental results. Small errors occur more often than large ones, and an error of given size will usually be equally often positive as negative. Such errors can be analysed statistically in terms of the distribution of error about a mean value; this distribution is a curve as in Fig. 7.1, which shows the number of times a particular error occurs as a function of the size of that error. Such a curve could be drawn if a large enough number of repeated measurements were made.

The simplest case is the one already mentioned, when errors are of random sign. The curve is then symmetrical about a maximum, which is the arithmetic mean. This is the normal, or Gaussian distribution of error, discussed in Section 7.7. If significant systematic error is present in addition to such random error, the mean will not be the true value and the curve may not be symmetrical.

Fig. 7.1 shows the probability that a particular result will be obtained in the measurement of a quantity x. We do not therefore expect repeated measurements to give the same result, and the quality of a set of measurements is assessed by the spread of results obtained. The uncertainty in the final result needs to be known and the conventional way of expressing that uncertainty is discussed in the next section.

Since the methods used to assess the size of experimental error are statistical, the sample size, which means the number of independent measurements made, is very important. The statistical theory is developed by assuming at first an infinitely large number of measurements. If we then add the results together, we sum over such a large number that the differences between adjacent results becomes so small that we can replace sums by integrals. For this reason sums and integrals occur in this chapter, depending on whether or not we are assuming an infinite number of measurements.

7.2   Root-mean-square error

If the measurement of a quantity x is repeated n times, the results will scatter about the arithmetic mean $\bar{x}$ given by

$$\bar{x} = \frac{\sum x_i}{n}, \tag{7.1}$$

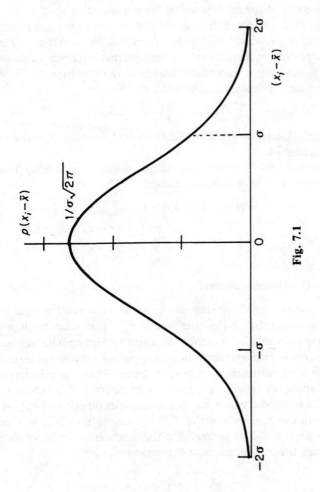

Fig. 7.1

and this is the best result obtainable from the measurements. The experimental error in a particular result x_i is then the difference between that value and the arithmetic mean, $(x_i - \bar{x})$. The final result of the experiment is to be expressed as $\bar{x} \pm \Delta x$, where Δx is a conventional assessment of the uncertainty of the measurement $\bar{x}$ of x.

In order that Δx shall be a useful measure of uncertainty it should reflect the spread in the values of x_i obtained. We cannot use for this the average value of the error $(x_i - \bar{x})$ because that is zero by definition of $\bar{x}$. We therefore use instead the square of the deviation, $(x_i - \bar{x})^2$, and define the root-mean-square error, $s(x)$, by

$$s(x) = \sqrt{[\sum (x_i - \bar{x})^2/n]}. \tag{7.2}$$

The relation between this and the standard deviation $\sigma(x)$ is discussed in Section 7.6.

The root-mean-square error can be calculated most easily from a set of values of x by use of the relation

$$\begin{aligned}
s(x)^2 = (1/n) \sum (x_i - \bar{x})^2 &= (1/n) \sum (x_i^2 - 2x_i\bar{x} + \bar{x}^2) \\
&= \sum x_i^2/n - 2\bar{x} \sum x_i/n + \bar{x}^2 \\
&= \sum x_i^2/n - \bar{x}^2.
\end{aligned} \tag{7.3}$$

7.3   Distribution of error

For discrete variables, such as the number of matches in a box, we define the probability $p(x)$ of obtaining a particular result x as the fraction of a very large number of samples which contain that number. In physical measurements, however, we are usually concerned with continuous, rather than discrete, variables. We can then define only the probability of obtaining a result in an interval of x between x and $x + dx$ as $p(x)dx$, where $p(x)$ is a continuous function of x. This is the smooth curve shown in Fig. 7.1. We ensure that $p(x)$ is a fraction between 0 and 1 by normalizing the function to unity, which means making the total area under the curve unity by

$$\int_{-\infty}^{+\infty} p(x)dx = 1. \tag{7.4}$$

The mean value $\bar{x}$ of x is obtained by summing (integrating) the product of x and its probability $p(x)$ over all possible values of x,

$$\bar{x} = \int_{-\infty}^{+\infty} xp(x)dx. \tag{7.5}$$

The root-mean-square error $\sigma(x)$ in x is then given by

$$\sigma(x)^2 = \int_{-\infty}^{+\infty} (x - \bar{x})^2 p(x)\mathrm{d}x. \tag{7.6}$$

7.4 The statistical analysis of experimental data

The methods used in the statistical analysis of data can be classified either from the mathematical or from the physical points of view, both of which will now be considered.

The simplest case is when measurements of a single quantity are repeatedly made. The results will scatter and, if a sufficiently large number of readings is taken, the distribution curve can be drawn. Only a small number of results will be available in real cases and there are standard tests, such as the 'chi-squared' test not considered here, to determine how closely a set of results follows a particular distribution. This has value as a way of testing for systematic error of a kind that distorts the distribution.

We are more often concerned with measurements of more than one variable. Statistical methods can then be used either to test whether or not the variables depend on each other, or to find the most probable values of numerical coefficients in an equation that is assumed to fit the data. The latter application is the more important one in physical science. From the statistical point of view there is no difference between these two applications because the uncertainty in the coefficients of a fitting equation can be regarded as a measure of the degree to which the assumed relation fits the data. Statistical texts discuss the concepts of covariance and correlation coefficient (both considered here) and of tests of significance and confidence limits (not considered) as measures of the extent to which variables are related rather than as measures of the uncertainty in the values of the fitting parameters.

Another way of comparing the various aspects of statistical analysis is in terms of the relevant number of independent variables. This is a physical concept, and for well defined chemical systems is easily determined. So long as all independent variables are identified and their values controlled in the experiment, a true value of any experimental quantity will exist, and results will be subject only to experimental error. On the other hand, properties and materials may not be rigorously defined and the independent variables may not be identifiable and controllable. A true value will not then exist and relations are being sought between quantities which both show legitimate variations, and this form of correlation analysis is not considered here.

7.5 Propagation of error

Experimental results are often substituted into equations so as to obtain values of related quantities. Knowing the uncertainty in the results we then need to calculate the uncertainty in the related quantity, for which we use the formula for propagation of error.

Suppose a quantity $z(x, y)$ depends on experimental variables x and y, and that measurements are repeated n times to obtain pairs of results $(x_1, y_1), (x_2, y_2), \ldots, (x_n, y_n)$. We define the probability of obtaining a result in the range $(x$ to $x + dx, y$ to $y + dy)$ as $p(x, y)dxdy$. The arithmetic mean value $\bar{z}$ of z is then

$$\bar{z} = \int_{-\infty}^{+\infty} \int_{-\infty}^{+\infty} z(x, y)p(x, y)dxdy. \tag{7.7}$$

We can also write the mean value of x in this form,

$$\bar{x} = \int_{-\infty}^{+\infty} \int_{-\infty}^{+\infty} xp(x, y)dxdy \tag{7.8}$$

because the integration with respect to y is simply the integral of the probability of obtaining any value of y for a particular value of x, which is normalized to unity by the two-dimensional analogue of (7.4)

The root-mean-square errors $\sigma(x)$, $\sigma(y)$ and $\sigma(z)$ in x, y and z respectively are defined as

$$\sigma(x)^2 = \int_{-\infty}^{+\infty} \int_{-\infty}^{+\infty} (x - \bar{x})^2 p(x, y)dxdy, \tag{7.9}$$

$$\sigma(y)^2 = \int_{-\infty}^{+\infty} \int_{-\infty}^{+\infty} (y - \bar{y})^2 p(x, y)dxdy, \tag{7.10}$$

$$\sigma(z)^2 = \int_{-\infty}^{+\infty} \int_{-\infty}^{+\infty} (z - \bar{z})^2 p(x, y)dxdy, \tag{7.11}$$

and we require an expression for $\sigma(z)$ in terms of $\sigma(x)$ and $\sigma(y)$.

Assuming an infinite number of measurements, the mean values will be free from experimental error so that $\bar{z}$ is obtained from $\bar{x}$ and $\bar{y}$. We use Taylor's theorem (Section 3.8) to express a departure of z from the arithmetic mean in the form

$$(z - \bar{z}) = \left(\frac{\partial z}{\partial x}\right)_y (x - \bar{x}) + \left(\frac{\partial z}{\partial y}\right)_x (y - \bar{y}) + \frac{1}{2}\left[\left(\frac{\partial^2 z}{\partial x^2}\right)(x - \bar{x})^2\right.$$
$$\left. + 2\left(\frac{\partial^2 z}{\partial x \partial y}\right)(x - \bar{x})(y - \bar{y}) + \left(\frac{\partial^2 z}{\partial y^2}\right)(y - \bar{y})^2\right]$$
$$+ \text{higher-order terms.} \tag{7.12}$$

We assume the errors to be small so that only the first-order terms in the Taylor expansion are significant. Then

$$(z - \bar{z})^2 = \left(\frac{\partial z}{\partial x}\right)^2 (x - \bar{x})^2 + \left(\frac{\partial z}{\partial y}\right)^2 (y - \bar{y})^2 + 2\left(\frac{\partial z}{\partial x}\right)\left(\frac{\partial z}{\partial y}\right)(x - \bar{x})(y - \bar{y}).$$
(7.13)

Substituting (7.13) into (7.11), assuming higher derivatives to be negligible so that the first derivatives are constants to be taken outside the integrals, we obtain

$$\sigma(z)^2 = \left(\frac{\partial z}{\partial x}\right)^2 \sigma(x)^2 + \left(\frac{\partial z}{\partial y}\right)^2 \sigma(y)^2 + 2\left(\frac{\partial z}{\partial x}\right)\left(\frac{\partial z}{\partial y}\right)\sigma(x, y) \quad (7.14)$$

where

$$\sigma(x, y) = \int_{-\infty}^{+\infty} \int_{-\infty}^{+\infty} (x - \bar{x})(y - \bar{y})p(x, y)\mathrm{d}x\mathrm{d}y. \quad (7.15)$$

The mean squares of the errors $\sigma(x)^2$ and $\sigma(y)^2$ are called the variances in x and y, and the new term $\sigma(x, y)$ is called the covariance of x and y. The significance of covariance is that it determines whether or not the variables are independent of each other. If they are independent, the integral in (7.15) separates into the product of two integrals each with respect to one variable only, and both of these integrals are zero by definition of the arithmetic mean. The covariance is therefore zero if the variables are independent.

In general, if z is a function of independent variables $x, y, u, \ldots$ the formula for propagation of error is

$$\sigma(z)^2 = \left(\frac{\partial z}{\partial x}\right)^2 \sigma(x)^2 + \left(\frac{\partial z}{\partial y}\right)^2 \sigma(y)^2 + \left(\frac{\partial z}{\partial u}\right)^2 \sigma(u)^2 + \ldots \quad (7.16)$$

7.6 Small-sample errors

We assume that a particular distribution curve such as Fig. 7.1 exists for the results of any experimental measurement. This curve is characterized by the form of the distribution, the arithmetic mean and the root-mean-square error. A particular form of distribution is considered in the next section, before which we consider relations that apply to any form of distribution.

The arithmetic mean has been defined for a series of n results by equation (7.1) and the corresponding root-mean-square error as $s(x)$ by equation (7.2). These have also been defined in terms of integrals by (7.5) and (7.6). These become the same only as the number of results tends to infinity because we require, in principle, an infinite number of points to establish the curve of distribution of error with complete certainty. The mean $\bar{x}_n$ of a sample of n results is only an estimate of the mean $\bar{x}_\infty$ of the distribution, so that a different sample of points will give a different value of $\bar{x}_n$; only as n tends to infinity will the means of different samples become the same and equal to the mean of the distribution. In the same way, the root-mean-square error of a sample of n results, denoted by $s(x)$, is only an estimate of that of the distribution $\sigma(x)$, which is called the standard deviation.

If we obtain n results and calculate the mean $\bar{x}_n$, and then repeat the set of n measurements a large enough number of times, we obtain a set of values of $\bar{x}_n$ which will themselves follow a distribution about the true mean $\bar{x}_\infty$. We can then calculate the standard deviation of the mean $\sigma(\bar{x}_n)$, which is a measure of the uncertainty in $\bar{x}_n$ due to the fact that we have too few results. By definition

$$\sigma(\bar{x}_n)^2 = \int_{-\infty}^{+\infty} (\bar{x}_n - \bar{x}_\infty)^2 p(x) \mathrm{d}x,$$

and since

$$\bar{x}_n = \frac{x_1}{n} + \frac{x_2}{n} + \ldots$$

the integral separates into the sum of n integrals of the form

$$\sigma(\bar{x}_n)^2 = \sum_{i=1}^{n} \frac{1}{n^2} \int_{-\infty}^{+\infty} (x_i - \bar{x}_\infty)^2 p(x) \mathrm{d}x = \sum_{i=1}^{n} \frac{\sigma(x)^2}{n^2} = \frac{\sigma(x)^2}{n},$$

so that

$$\sigma(\bar{x}_n) = \frac{\sigma(x)}{\sqrt{n}}. \tag{7.17}$$

This relation shows that when n duplicate measurements are made of the same quantity, each of which is subject to error with standard deviation $\sigma(x)$, and the mean is calculated so as to obtain a more reliable estimate, the uncertainty is thereby reduced by the factor $1/\sqrt{n}$. Thus the mean of nine duplicate readings will reduce the uncertainty by a factor of one-third, but to improve the precision by one decimal place we would need to take 100 readings. It is seldom worth while to repeat a reading more than about 10 times, except when automatic recording is available. When it is possible to take readings very quickly, automatic repetition is especially useful because the effect of even high-frequency noise can then be averaged out.

The effect of small-sample error on the estimate $s(x)$ of the standard deviation can also be calculated. Here we are seeking not the error in the measurement, but the error in our estimate of that error, which is of second order in small quantities. If the true mean of the distribution, $\bar{x}_\infty$, were known we could obtain a good estimate of the standard deviation from only n results by using the relation

$$\sigma(x)^2 = \frac{\sum (x_i - \bar{x}_\infty)^2}{n}. \tag{7.18}$$

We introduce $\bar{x}_n$ into this expression by writing the algebraic identity

$$(x_i - \bar{x}_\infty) = (x_i - \bar{x}_n) - (\bar{x}_n - \bar{x}_\infty)$$

so that

$$(x_i - \bar{x}_\infty)^2 = (x_i - \bar{x}_n)^2 - 2(x_i - \bar{x}_n)(\bar{x}_n - \bar{x}_\infty) + (\bar{x}_n - \bar{x}_\infty)^2,$$

and

$$\sum (x_i - \bar{x}_\infty)^2 = \sum (x_i - \bar{x}_n)^2 + n(\bar{x}_n - \bar{x}_\infty)^2, \tag{7.19}$$

the middle term having disappeared because $\sum (x_i - \bar{x}_n) = 0$ for the arithmetic mean. We can then write the required relation between $s(x)$ and $\sigma(x)$; by definition

$$s(x)^2 = \frac{\sum (x_i - \bar{x}_n)^2}{n}, \tag{7.20}$$

and combining (7.18), (7.19) and (7.20) gives

$$n\sigma(x)^2 = \sum (x_i - \bar{x}_n)^2 + n(\bar{x}_n - \bar{x}_\infty)^2$$
$$= ns(x)^2 + n\sigma(\bar{x}_n)^2$$
$$= ns(x)^2 + \sigma(x)^2$$

so that

$$\sigma(x)^2 = \frac{n}{n-1} s(x)^2 = \frac{\sum (x_i - \bar{x}_n)^2}{n-1}, \tag{7.21}$$

which is the equation for the standard deviation of a set of measurements of a quantity x. The estimate $s(x)$ of the standard deviation $\sigma(x)$ is said to be biased by the smallness of the sample, this bias being corrected by the factor $n/(n-1)$. This can be interpreted by noting that the calculation of the mean involves the use of an equation containing the n variables, so that $(n-1)$ variables remain independent, or we have $(n-1)$ degrees of freedom. By analogy, if we fit n sets of values of several experimental variables to an equation containing k numerical coefficients, we then have $(n-k)$ degrees of freedom and the standard deviation of the fit is

$$\sigma(z)^2 = \frac{\sum (z_i - z_{calc})^2}{n-k} \tag{7.22}$$

where $(z_i - z_{calc})$ is the deviation of an experimental value of z from the fitting equation.

7.7 The normal distribution of error

We assume random sign error in a set of independent measurements of an experimental variable x. This means that errors of given magnitude occur equally often with positive as with negative sign, and the sum of the errors is zero. When error is defined as departure from a mean value m, this must be the arithmetic mean because

$$\sum (x_i - m) = \sum x_i - nm = 0$$

so that

$$m = \sum x_i / n = \bar{x}.$$

We can find an equation for the probability $p(x_i - m)$ of obtaining error $(x_i - m)$ by maximizing the probability of obtaining a set of independent errors subject to the condition that m is the arithmetic mean.

The probability P of obtaining a set of results $x_1, x_2, \ldots, x_i, \ldots$ is the product of the probabilities of obtaining each result,

$$P = \prod p(x_i - m)$$

which will be a maximum when $dP/dm = 0$, or

$$\frac{d \ln P}{dm} = \frac{1}{P} \frac{dP}{dm} = 0.$$

Introducing the condition that m is the arithmetic mean $\bar{x}$, we have

$$\sum (x_i - \bar{x}) = 0 \tag{7.23}$$

and

$$\frac{d \ln P}{dm} = \sum \frac{d \ln p(x_i - m)}{dm} = \sum \frac{d \ln p(x_i - \bar{x})}{d\bar{x}} = 0,$$

so that

$$\sum [d \ln p(x_i - \bar{x})/d(x_i - \bar{x})] = 0. \tag{7.24}$$

Solving equations (7.23) and (7.24) simultaneously by adding λ times equation (7.23) to equation (7.24) (compare Section 2.8) we obtain

$$\frac{d \ln p(x_1 - \bar{x})}{d(x_1 - \bar{x})} + \lambda(x_1 - \bar{x}) + \frac{d \ln p(x_2 - \bar{x})}{d(x_2 - \bar{x})} + \lambda(x_2 - \bar{x}) + \ldots = 0. \tag{7.25}$$

Hence, by suitable choice of λ,

$$\mathrm{d}\ln p(x_i - \bar{x}) = -\lambda(x_i - \bar{x})\mathrm{d}(x_i - \bar{x})$$

and integration gives

$$p(x_i - \bar{x}) = K \exp[-\lambda(x_i - \bar{x})^2/2].$$

The integration constant K is determined by normalizing $p(x_i - \bar{x})$ to unity, so that

$$\int_{-\infty}^{+\infty} p(x_i - \bar{x})\mathrm{d}(x_i - \bar{x}) = 1,$$

or since $p(x_i - \bar{x})$ is symmetrical about $x_i = \bar{x}$,

$$\int_{0}^{+\infty} p(x_i - \bar{x})\mathrm{d}(x_i - \bar{x}) = \tfrac{1}{2}.$$

This gives an integral of the standard form discussed in Example 5.1, from which

$$K\int_{0}^{+\infty} \exp[-\lambda(x_i - \bar{x})^2/2]\mathrm{d}(x_i - \bar{x}) = K\sqrt{(\pi/2\lambda)} = \tfrac{1}{2}$$

$$K = \sqrt{(\lambda/2\pi)}$$

and

$$p(x_i - \bar{x}) = \sqrt{(\lambda/2\pi)}\exp[-\lambda(x_i - \bar{x})^2/2].$$

Writing the first and second derivatives of $p(x_i - \bar{x})$ with respect to $(x_i - \bar{x})$,

$$\mathrm{d}p(x_i - \bar{x})/\mathrm{d}(x_i - \bar{x}) = -\sqrt{(\lambda/2\pi)}\lambda(x_i - \bar{x})\exp[-\lambda(x_i - \bar{x})^2/2]$$
$$\mathrm{d}^2p(x_i - \bar{x})/\mathrm{d}(x_i - \bar{x})^2 = \sqrt{(\lambda/2\pi)}\lambda[\lambda(x_i - \bar{x})^2 - 1]\exp[-\lambda(x_i - \bar{x})^2/2].$$

At $x_i = \bar{x}$ the first derivative is zero and the second is negative, so that $p(x_i - \bar{x})$ is a maximum at the arithmetic mean, as required. The curve will have the shape shown in Fig. 7.1, and the halfwidth of the curve can be expressed as the error at the point of inflection; this is when the second derivative is zero, or

$$\lambda(x_i - \bar{x})^2 - 1 = 0$$

$$(x_i - \bar{x}) = \pm 1/\sqrt{\lambda} = \pm\sigma, \quad \text{say.}$$

Then

$$p(x_i - \bar{x}) = \frac{1}{\sigma\sqrt{(2\pi)}}\exp[-(x_i - \bar{x})^2/2\sigma^2] \qquad (7.26)$$

and σ is called the standard deviation. This can be identified with the root-mean-square error $\sigma(x)$ by

$$\sigma(x)^2 = \int_{-\infty}^{+\infty} (x_i - \bar{x})^2 p(x_i - \bar{x}) \mathrm{d}(x_i - \bar{x})$$

$$= 2 \int_0^\infty (x_i - \bar{x})^2 \{1/[\sigma\sqrt{(2\pi)}]\} \exp[-(x_i - \bar{x})^2/2\sigma^2] \mathrm{d}(x_i - \bar{x})$$

$$= \sigma^2 \qquad \text{by Example 5.2.}$$

Equation (7.26) describes the normal or Gaussian distribution of error. We can expect experimental measurements to follow this distribution so long as only random sign errors are present.

A theorem that is important in practice is the central limit theorem, which states that if mean values are taken of samples having any distribution, those means follow a normal distribution. Such averaging can occur in the use of measuring instruments, so that we then obtain a normal distribution of error even when the quantity being measured does not itself follow that distribution.

7.8   The method of least squares

Given a normal distribution of error, the probability of obtaining a set of results $x_1, x_2, \ldots, x_i$ is

$$P = \prod[p(x_i - \bar{x})]$$

$$= \left(\frac{1}{\sigma\sqrt{(2\pi)}}\right)^n \exp\left(\frac{-\sum(x_i - \bar{x})^2}{2\sigma^2}\right). \qquad (7.27)$$

This will be a maximum when

$$\sum(x_i - \bar{x})^2 \quad \text{is a minimum.}$$

This is the principle of least squares, that the sum of the squares of the deviations is a minimum.

7.8.1   Linear relation between two variables

Given that experimental variables x and y are related by the equation

$$y = mx + c, \qquad (7.28)$$

we wish to obtain the most probable values of m and c to fit a set of independently measured points $(x_1, y_1), (x_2, y_2), \ldots, (x_i, y_i), \ldots, (x_n, y_n)$. In general, both x and y will be in error. We assume a normal distribution of error for each variable at each point, the means of which distributions lie on the straight line

$$\bar{y}_i = m\bar{x}_i + c. \tag{7.29}$$

In Fig. 7.2, curves illustrating the distributions parallel to the x- and the y-axes have been drawn displaced for clarity; these curves should be interpreted as being centred on $(\bar{x}_i, \bar{y}_i)$ and in the planes perpendicular to the paper and parallel to the axes. An error in x displaces the point from the centre of its distribution to, say, the position of the triangle, and error in y to the position of the square. The observed point is then found at X.

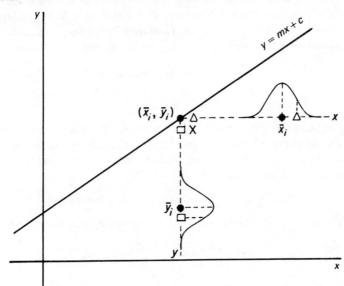

Fig. 7.2

When we assume independent random errors in x and in y, the probability of obtaining point (x_i, y_i) is the product of the probabilities $p(x_i - \bar{x}_i)$ and $p(y_i - \bar{y}_i)$. If we assume the standard deviation $\sigma(x)$ to be the same at each of the n points, and similarly for $\sigma(y)$, we obtain

$$p(x_i - \bar{x}_i, y_i - \bar{y}_i) = p(x_i - \bar{x}_i)p(y_i - \bar{y}_i),$$

so that

$$P = \left(\frac{1}{2\pi\sigma(x)\sigma(y)}\right)^n \exp\left[-\left(\frac{\sum(x_i - \bar{x}_i)^2}{2\sigma(x)^2} + \frac{\sum(y_i - \bar{y}_i)^2}{2\sigma(y)^2}\right)\right],$$

which will be a maximum when

$$\sigma(y)^2 \sum(x_i - \bar{x}_i)^2 + \sigma(x)^2 \sum(y_i - \bar{y}_i)^2 \quad \text{is a minimum.} \tag{7.30}$$

Further assumptions are necessary for the best line to be found because so far we have more unknowns than data points.

If a group of n points from a chosen range of x is taken and the mean $(\bar{x}, \bar{y})$ is calculated, this mean point will then have a smaller uncertainty than the individual points by the factor $1/\sqrt{n}$ (equation (7.17)). A recommended procedure when the errors in both x and y are significant is, therefore, to divide the points into groups at the upper and lower regions of the variables, calculate the mean point of each group, and use these two points to define the straight line.

In real applications the errors in the two variables are often not the same. We then assume the better known variable to be free from error, this being used as the independent variable x. When we assume error to be vested only in y we have $x_i = \bar{x}_i$, and condition (7.30) becomes

$$\sum (y_i - \bar{y}_i)^2 \quad \text{is a minimum}$$

with

$$\bar{y}_i = mx_i + c.$$

We put $d_i = y_i - mx_i - c$ and call this the residual; this is the deviation of the point (x_i, y_i) from the line measured parallel to the y-axis. The least-squares condition is then that the sum of the squares of the residuals shall be a minimum,

$$D = \sum d_i^2 = \sum (y_i - mx_i - c)^2 \quad \text{is a minimum.}$$

We now regard D as a function of the two unknowns m and c, so that

$$dD = \left(\frac{\partial D}{\partial m}\right) dm + \left(\frac{\partial D}{\partial c}\right) dc = 0.$$

A solution of this equation will be

$$\left(\frac{\partial D}{\partial m}\right) = \left(\frac{\partial D}{\partial c}\right) = 0$$

giving

$$\frac{\partial}{\partial m} \sum (y_i - mx_i - c)^2 = -2\sum x_i(y_i - mx_i - c) = 0,$$

and since we may now omit the subscript i without causing confusion,

$$\sum xy - m\sum x^2 - c\sum x = 0, \tag{7.31}$$

$$\frac{\partial}{\partial c} \sum (y_i - mx_i - c)^2 = -2\sum (y_i - mx_i - c) = 0,$$

$$\sum y - m\sum x - nc = 0. \tag{7.32}$$

Solving (7.31) and (7.32) simultaneously for m and c gives the equivalent forms of equation

$$m = \frac{n\sum xy - \sum x \sum y}{n\sum x^2 - (\sum x)^2} = \frac{\sum(x - \bar{x})(y - \bar{y})}{\sum(x - \bar{x})^2} = \frac{\sum(x - \bar{x})y}{\sum(x - \bar{x})^2} \quad (7.33)$$

$$c = \frac{\sum y \sum x^2 - \sum x \sum xy}{n\sum x^2 - (\sum x)^2} = \bar{y} - m\bar{x}. \quad (7.34)$$

Equation (7.34) shows that the least-squares line passes through the 'centre of gravity' of the points, $(\bar{x}, \bar{y})$, and since (7.32) may be written as

$$\sum(y_i - mx_i - c) = 0$$

we see that minimizing the sum of the squares of the residuals incidentally makes the simple sum of the residuals also equal to zero.

7.8.2 Covariance and correlation coefficient

The covariance $\sigma(x, y)$ of variables x and y is defined in Section 7.5 and shown to be zero if x and y are independent. On the other hand, if x and y are linearly dependent, the slope m of the least-squares line $y = mx + c$ can be expressed in terms of $\sigma(x, y)$. For a finite number of points, m is given by (7.33) and the estimates $s(x, y)$ of $\sigma(x, y)$, $s(x)^2$ of $\sigma(x)^2$ and $s(y)^2$ of $\sigma(y)^2$ are

$$s(x, y) = [\sum(x - \bar{x})(y - \bar{y})]/n, \quad (7.35)$$
$$s(x)^2 = [\sum(x - \bar{x})^2]/n, \quad (7.36)$$
$$s(y)^2 = [\sum(y - \bar{y})^2]/n, \quad (7.37)$$

so that the estimated value of m is

$$m = s(x, y)/s(x)^2. \quad (7.38)$$

The values of m and c for the least-squares line will be uncertain because only a limited number of points are available. As the number of points tends to infinity the uncertainty in the line tends to zero, $s(x, y) \to \sigma(x, y)$, $s(x)^2 \to \sigma(x)^2$ and $\sum(y - \bar{y}) \to m\sum(x - \bar{x})$, so that

$$s(y)^2 = \frac{\sum(y - \bar{y})^2}{n} = \frac{m^2\sum(x - \bar{x})^2}{n} = m^2 s(x)^2.$$

Hence for an infinite number of points

$$\sigma(y)^2 = m^2 \sigma(x)^2$$

and (7.38) gives

$$\sigma(x, y) = m\sigma(x)^2 = \sigma(x)\sigma(y). \quad (7.39)$$

The ratio

$$\rho = \frac{\sigma(x, y)}{\sigma(x)\sigma(y)} \tag{7.40}$$

is called the correlation coefficient. This will be zero if x and y are independent since $\sigma(x, y)$ will then be zero. When x and y are linearly related, equations (7.39) and (7.40) show that $\rho = 1$, and this provides a test for such a linear relation.

When this is applied to a finite number of points we can calculate only an estimate $r(x, y)$ of the correlation coefficient, given by

$$r = r(x, y) = s(x, y)/s(x)s(y), \tag{7.41}$$

which may be written in the alternative forms

$$
\begin{aligned}
r^2 &= \frac{[\sum(x - \bar{x})(y - \bar{y})]^2}{\sum(x - \bar{x})^2 \sum(y - \bar{y})^2} \\
&= \frac{nc\sum y + nm\sum xy - (\sum y)^2}{n\sum y^2 - (\sum y)^2} \\
&= \frac{(n\sum xy - \sum x\sum y)^2}{[n\sum x^2 - (\sum x)^2][n\sum y^2 - (\sum y)^2]}.
\end{aligned} \tag{7.42}
$$

This will not be unity even when a true straight line exists because small-sample error means that expressions (7.35)–(7.37) are not the same as (7.9), (7.10) and (7.15). A value of r that is less than unity is therefore evidence either of limitations in the number and precision of the points, or that the supposed linear relation does not apply; it is not possible to separate these two possibilities for a given set of results.

A correlation coefficient that is much less than unity does not imply independence of the two variables since it may be that a curve rather than a straight line fits the data. This may be seen by plotting the graph, and the variables should then be changed to convert a curve into a straight line before applying the least-squares analysis.

7.8.3   Uncertainty in the slope and intercept of the least-squares straight line

Equations (7.33) and (7.34) for m and c were derived assuming the errors to be vested entirely in the variable y. The error δy_i is the difference between y_i and the value $mx_i + c$ given by the line,

$$\delta y_i = y_i - mx_i - c. \tag{7.43}$$

From equation (7.33)

$$m = \frac{\sum (x_i - \bar{x}) y_i}{\sum (x_i - \bar{x})^2}$$

and by propagation of error (Section 7.5), with $(x_i - \bar{x})$ assumed to be precisely known,

$$(\delta m)^2 = \sum \left[\left(\frac{\partial m}{\partial y_i} \right)^2 \delta y_i^2 \right].$$

Writing the mean value of δy_i^2 as σ_y^2, and assuming this to be the same for all y_i,

$$(\delta m)^2 = \frac{\sum (x_i - \bar{x})^2 \sigma_y^2}{[\sum (x_i - \bar{x})^2]^2} = \frac{\sigma_y^2}{\sum (x_i - \bar{x})^2} = \frac{n \sigma_y^2}{n \sum x^2 - (\sum x)^2} \quad (7.44)$$

where

$$\sigma_y^2 = \frac{\sum (y_i - m x_i - c)^2}{(n-2)} = \frac{(1 - r^2)[n \sum y^2 - (\sum y)^2]}{n(n-2)}. \quad (7.45)$$

The divisor $(n-2)$ in (7.45) arises because two numerical coefficients, m and c, are derived from the data leaving $(n-2)$ degrees of freedom.

When the least-squares line is used to estimate the value y_0 of y corresponding to the value x_0 of x we have

$$y_0 = \bar{y} + m(x_0 - \bar{x})$$

so that

$$(\delta y_0)^2 = (\delta \bar{y})^2 + (x_0 - \bar{x})^2 (\delta m)^2.$$

Again assuming a mean value σ_y^2 for all values of y, the error in the mean value of y is given by equation (7.17)

$$(\delta \bar{y})^2 = \sigma_y^2 / n$$

which, together with (7.44), gives

$$(\delta y_0)^2 = \sigma_y^2 \left(\frac{1}{n} + \frac{(x_0 - \bar{x})^2}{\sum (x_i - \bar{x})^2} \right) = \frac{\sigma_y^2}{n} + \frac{(x_0 - \bar{x})^2 n \sigma_y^2}{n \sum x^2 - (\sum x)^2} \quad (7.46)$$

The uncertainty in the intercept c is then found by putting $x_0 = 0$;

$$(\delta c)^2 = \sigma_y^2 \left(\frac{1}{n} + \frac{\bar{x}^2}{\sum (x_i - \bar{x})^2} \right)$$

$$= \frac{\sigma_y^2}{n} \left(1 + \frac{(\sum x)^2}{n \sum x^2 - (\sum x)^2} \right) \quad (7.47)$$

with σ_y^2 given by equation (7.45). By combining equations (7.44) and

(7.47) we obtain

$$(\delta c)^2 = \frac{\sigma_y^2}{n} + \bar{x}^2 (\delta m)^2.$$

Example 7.1

The result of applying this analysis to a set of points is shown in Fig. 7.3. The mean point $(\bar{x}, \bar{y})$ has error bars at $\pm \sigma_y / \sqrt{n}$. The least-squares line, a, is drawn with alternative slopes (b and c) of $\pm \delta m$. The curves d and e are drawn to show the errors $\pm \delta y_0$ at particular values of x_0 as given by equation (7.46). The error $\pm \delta c$ in the intercept on the y-axis is due mainly to the second term in the bracket in (7.47), corresponding to uncertainty in the slope, with an additional contribution from the first term which corresponds to the uncertainty in $\bar{y}$. The correlation coefficient for the points shown is $r = 0.775$; it can be seen that as low a value as this indicates a considerable scatter in the points.

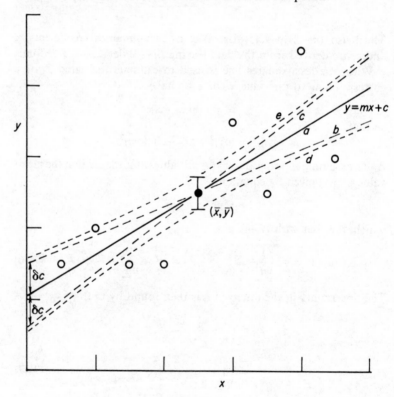

Fig. 7.3

7.8.4 *Least-squares straight line with both variables subject to error*

The calculation of the least-squares line in Section 7.8.1 assumes that the error is vested in only one of the two variables and equations (7.33) and (7.34) for m and c assume the error to be in y. Experimental conditions may justify this assumption, otherwise we might equally well assume the error to be vested only in x; we would then obtain a different straight line by interchanging x and y in the equations.

Since the least-squares line passes through the 'centre of gravity' of the points, the two lines obtained by assuming the error to be vested either in y or in x will intersect at that point. The ratio of the slopes of these lines is related to the estimate r of the correlation coefficient. From equation (7.38)

$$m = \frac{s(x, y)}{s(x)^2} = \frac{\sum (x - \bar{x})(y - \bar{y})}{\sum (x - \bar{x})^2}. \tag{7.48}$$

We write the alternative line as $x = m'y + c'$ and by interchanging x and y we obtain

$$m' = \frac{s(x, y)}{s(y)^2} = \frac{\sum (x - \bar{x})(y - \bar{y})}{\sum (y - \bar{y})^2}. \tag{7.49}$$

Now m is the slope dy/dx, whereas m' is dx/dy; we therefore compare m with $1/m'$. The ratio of the slopes is then

$$\frac{m}{1/m'} = mm' = \frac{s(x, y)^2}{s(x)^2 s(y)^2} = r^2. \tag{7.50}$$

As the number of points tends to infinity, the estimated correlation coefficient r tends to the value ρ free from small-sample error. The alternative lines will then move closer together as the number of points is increased, and will coincide for an infinite number of points if the linear relation applies so that $\rho = 1$. It is conventional to assume that if the errors in the two variables are both significant the best fit to the data passes through $(\bar{x}, \bar{y})$ and bisects the angle between the alternative lines, with slope equal to the mean of m and $1/m'$.

7.8.5 *Weighting of observations*

We have assumed so far that all experimental points are equally reliable, having used the same uncertainty σ_y for all values of y. There may, however, be experimental reasons for regarding some points as more reliable than others; this can be included in the analysis by weighting the points. The simplest way of achieving this is to repeat the entry of any particular point into the analysis w times, thus making that point w times as significant as a single point. It is necessary to remember only that these weighting factors must be included in an effective total number of points.

The proper weight to be used in precise work can be seen from the expression for the normal distribution, equation (7.27); the quantity to be minimized is the sum of the squares of the deviations divided by σ^2, so that each point should be weighted by the reciprocal of its standard deviation, such as might be estimated by the formula for propagation of error.

7.8.6　*Multivariable and non-linear least-squares analysis*

When fitting results to a more complicated equation than that of a straight line, the analytical methods of this chapter are still applicable and remain useful in simple cases. Assuming the error to be vested in a particular variable, the difference between an experimental value of the variable and that predicted by an assumed equation is the residual d_i. The principle of least squares is that $\sum(d_i^2)$ is a minimum. Partial differentiation of $\sum(d_i^2)$ with respect to each numerical coefficient in turn gives a set of normal equations, which can then be solved. When these equations are linear in each coefficient the solution is straightforward; otherwise linearization may be achieved by using trial values of the coefficients, for which a text on statistical analysis should be consulted.

An alternative to the analytical approach is to use trial-and-error searching by computer to find values of the coefficients that produce the smallest value of $\sum(d_i)^2$. Standard programs are available for this purpose, some of which make use of expressions for the partial derivatives with respect to the coefficients to accelerate convergence. It is important when using numerical methods to allow only the smallest possible number of search parameters; an excessive number will produce an undulating curve or surface which may fit the data precisely but be physically misleading. A polynomial fit in which the coefficients alternate in sign may be evidence of this effect.

SI units, physical constants and conversion factors; the Greek alphabet and a summary of useful relations

SI units

SI stands for the agreed international system of units, which is based upon the kilogram (kg), metre (m) and second (s), together with the kelvin (K) unit of temperature, the mole (mol) for amount of substance and the ampere (A). The candela (cd) unit of luminous intensity is also a basic unit. Other units are derived from these basic units.

The kelvin is defined as 1/273.16 of the thermodynamic temperature of the triple point of water. This means any subsequent refinements in absolute temperature measurement will alter the size of the degree and so, say, the absolute temperature of the boiling point of water, without changing the triple point. The ice point is 0.01 K below the triple point so that 0°C is 273.15 K.

The mole is the amount of substance containing as many elementary units as there are atoms in 0.012 kg of ^{12}C. This allows experimental measurement of the Avogadro constant by, in effect, counting this number of atoms, and of the gas constant from measurements of the properties of gases.

The unit of force is the newton (N), defined as that which produces an acceleration of 1 m s^{-2} in a mass of 1 kg. Pressure is then force per unit area in N m^{-2}, this unit being called the pascal (Pa). Energy is measured in joule (J), which is 1 N m, and power in watt (w) which is 1 J s^{-1}.

The ampere (A) is that current which produces a force of 2×10^{-7} N m^{-1} between long straight parallel conductors 1 m apart in vacuum. The volt (V) is then the potential difference such that a current of 1 A dissipates 1 watt, and the coulomb (C) is 1 A s. The faraday constant (F) can then be measured as the number of coulombs of charge carried by one avogadro of electrons.

Standard gravity is 9.80 665 m s^{-2} and a standard atmosphere is the pressure produced by a mercury column 0.76 m high under standard gravity at 0°C; this is 101.325 kPa. The preferred unit is the bar, defined as 100 kPa. Other units that are still in use are the calorie in various forms, usually the thermochemical calorie of 4.184 J, and the Angstrom, Å, which is 10^{-10} m. The word 'litre' is now regarded as a special name for the cubic decimetre, with the recommendation that neither the word litre nor its symbol, l, should be used to express results of high precision. This is because the litre was previously defined as 1.000 028 dm^3, this definition being rescinded in 1964.

Physical constants

Recommended values of the fundamental physical constants are published by the International Union of Pure and Applied Chemistry. Some of the values given in *Pure and Applied Chemistry*, **51**, 1 (1979) are:

Avogadro constant L	$6.022045(31) \times 10^{23}$ mol^{-1}
Boltzmann constant k	$1.380662(44) \times 10^{-23}$ J K^{-1}
gas constant R	$8.31441(26)$ J K^{-1} mol^{-1}
Planck constant h	$6.626176(36) \times 10^{-34}$ J s
Faraday constant F	$9.648456(27) \times 10^4$ C mol^{-1}
mass of proton	$1.6726485(86) \times 10^{-27}$ kg
charge on electron e	$1.6021892(46) \times 10^{-19}$ C
mass of electron	$9.109534(47) \times 10^{-31}$ kg
speed of light in vacuum	$2.99792458(1) \times 10^8$ m s^{-1}
permittivity of vacuum	$8.85418782(5) \times 10^{-12}$ C^2 N^{-1} m^{-2}
normal atmosphere	1.01325×10^5 Pa exactly
zero of Celsius scale	273.15 K exactly
standard acceleration of free fall	9.80665 m s^{-2} exactly

The notation used is that the figure(s) in brackets is (are) the standard deviation of the last figure(s) quoted.

Conversion factors for units

1 amp = 1 A = 1 J s^{-1} V^{-1}

1 atm = 101.325 kPa

1 cal = 4.184 J

1 coulomb = 1 C = 1 A s = 1 J V^{-1}

0.30 debye = 10^{-30} C m

1 dyne = 10^{-5} N

1 electron volt = 1 eV = 1.6021 × 10^{-19} J

1 erg = 10^{-7} J

1 farad = 1 F = 1 A s V^{-1}
$\qquad$ = 1 m^{-2} kg^{-1} s^4 A^2

1 gallon (UK) = 4.546 litre
$\qquad$ = 1.2009 gallon (USA)

1 inch = 2.540 cm

1 joule = 1 J = 1 N m

1 m^3 = 10^6 cm^3

1 micron = 1 μ = 1 μm = 10^{-6} m

1 newton = 1 N = 1 m kg s^{-2}

1 pascal = 1 Pa = 1 N m^{-2} = 1 m^{-1} kg s^{-2}

1 pound (UK) = 1 lb = 0.4536 kg

1 torr = 1 mmHg = 0.1333 kPa

1 volt = 1 V = 1 J A^{-1} s^{-1}
$\qquad$ = 1 m^2 kg s^{-3} A^{-1}

1 watt = 1 W = 1 J s^{-1} = 1 m^2 kg s^{-3}

The Greek alphabet

A	α	alpha	N	ν	nu
B	β	beta	Ξ	ξ	xi
Γ	γ	gamma	O	o	omicron
Δ	δ	delta	Π	π	pi
E	ε	epsilon	P	ρ	rho
Z	ζ	zeta	Σ	σ	sigma
H	η	eta	T	τ	tau
Θ	θ	theta	Y	u	upsilon
I	i	iota	Φ	ϕ	phi
K	κ	kappa	X	χ	chi
Λ	λ	lambda	Ψ	ψ	psi
M	μ	mu	Ω	ω	omega

Summary of useful relations

$\cos^2 A + \sin^2 A = 1$

$\cos^2 A - \sin^2 A = \cos 2A$

$\sin 2A = 2\sin A \cos A$

$\sin (A \pm B) = \sin A \cos B \pm \cos A \sin B$

$\cos (A \pm B) = \cos A \cos B \mp \sin A \sin B$

$e^x e^y = e^{x+y}$

$a^{-x} = 1/a^x$

if $y = e^x$ then $x = \ln y$

$\ln x = \log_e x = 2.3026 \log x = 2.3026 \log_{10} x$

$\log_{10} 10^{-x} = -x$

$$ax^2 + bx + c = 0; \quad x = -\frac{b}{2a} \pm \frac{1}{2a} \sqrt{(b^2 - 4ac)}$$

$$e^x = 1 + x + \frac{x^2}{2!} + \frac{x^3}{3!} + \dots$$

$$\ln (1 + x) = x - \frac{x^2}{2} + \frac{x^3}{3} - \dots$$

$$\sin x = x - \frac{x^3}{3!} + \frac{x^5}{5!} - \dots$$

$$\cos x = 1 - \frac{x^2}{2!} + \frac{x^4}{4!} - \dots$$

$$(1 + x)^n = 1 + nx + \frac{n(n-1)}{2!} x^2 + \dots$$

$$(a + x)^n = a^n + na^{n-1} x + \frac{n(n-1)}{2!} a^{n-2} x^2 + \dots$$

$$f(x) = f(0) + xf'(0) + \frac{x^2}{2!} f''(0) + \dots$$

$$\lim_{x \to a} \frac{f(x)}{g(x)} = \lim_{x \to a} \frac{f'(x)}{g'(x)} \quad \text{if } f(a) = g(a) = 0 \text{ or } \infty$$

$$z = z(x, y); \quad dz = \left(\frac{\partial z}{\partial x}\right)_y dx + \left(\frac{\partial z}{\partial y}\right)_x dy$$

$$\frac{dy}{dx} = \frac{dy}{du} \frac{du}{dx}$$

$$\left(\frac{\partial z}{\partial x}\right)_y = -\left(\frac{\partial z}{\partial y}\right)_x \left(\frac{\partial y}{\partial x}\right)_z$$

$$\int u dv = uv - \int v du$$

$$\left(\frac{\partial z}{\partial y}\right)_u = \left(\frac{\partial z}{\partial y}\right)_x + \left(\frac{\partial z}{\partial x}\right)_y \left(\frac{\partial x}{\partial y}\right)_u$$

$$e^{iy} = \cos y + i \sin y$$

$$e^{-iy} = \cos y - i \sin y$$

Index